AF396491

THE
PANZER
ARCHIVES
VOLUME 1

THE PANZER ARCHIVES

VOLUME 1

PANZER II–IV, PRAGA TNH-P (PANZER 38(t)), E-100, MOUSE 1-MAN TANK AND ARMOURED CARS

CRAIG MOORE

FONTHILL

Cover: Top photographs left and right: A Czechoslovakian-built Praga TNH-P 8-ton tank that was sent to Britain for evaluation in 1938. It was later called the Panzer 38(t) in German army service. Bottom photographs left and right: A captured German Panzer IV Ausf.D tank sent to the UK for examination.

First published in Great Britain in 2026 by
Fonthill
An imprint of
Pen & Sword Books Ltd
Yorkshire – Philadelphia
www.fonthill.media

Typeset in SabonLTStd 11/14 by
SJmagic DESIGN SERVICES, India.
Printed and bound in the UK by CPI Group (UK) Ltd, Croydon, CR0 4YY

The Publisher's authorised representative in the EU for product safety is Authorised Rep Compliance Ltd., Ground Floor, 71 Lower Baggot Street, Dublin D02 P593, Ireland.
www.arccompliance.com

For a complete list of Pen & Sword titles please contact
PEN & SWORD BOOKS LIMITED
George House, Units 12 & 13, Beevor Street, Off Pontefract Road,
Barnsley, South Yorkshire, S71 1HN, England
E-mail: enquiries@pen-and-sword.co.uk
Website: www.pen-and-sword.co.uk

or

PEN AND SWORD BOOKS
1950 Lawrence Rd, Havertown, PA 19083, USA
E-mail: uspen-and-sword@casematepublishers.com
Website: www.penandswordbooks.com

Acknowledgements

Herbert Ackermans, Jason Belgrave, Pierre-Oliver Buan, Lauren Child, Hilary Louis Doyle, Massimo Foti, Marcus Hock, Yuri Pasholok, Ed Webster.

Contents

Introduction

During the Second World War, the British War Office instructed their military engineers and scientists to examine captured enemy armoured vehicles and tanks to find out their mechanical specifications and abilities, but more importantly how to destroy them using Allied weaponry. This vital information was then passed on to the troops on the battlefields via briefings and technical summaries attached to General Orders.

After the war, these secret reports were sent for storage to different archives around Britain. They have now been de-classified. 'The Panzer Archives' book series brings those reports together in a single collection. Most have not been published before. The military style and formatting of the original text, spelling plus abbreviations have been kept. The notation 9' 2" that is found in some of the reports is the abbreviated form of feet and inches: 9' 2" = 9 feet 2 inches. The photographs contained in the reports have been reproduced. A file may contain reports from different departments dealing with the same vehicle. Sometimes information is repeated. This text has been kept in because it usually contains some different information not contained in previous reports. This book series enables the reader access to the information in these fascinating historical documents without having to go to the archives.

This volume contains the British findings and impressions of the captured German Panzer II tank, Panzer III tank, Panzer IV tank, German tank protective wire mesh skirting, E 100 tank, the Mouse one-man light tank, German light amphibious Turtle armoured cars, and the Italian AB41 armoured car, plus the Czechoslovakian-built Praga TNH-P 8-ton tank that would later see service in the German army as the Panzer 38(t).

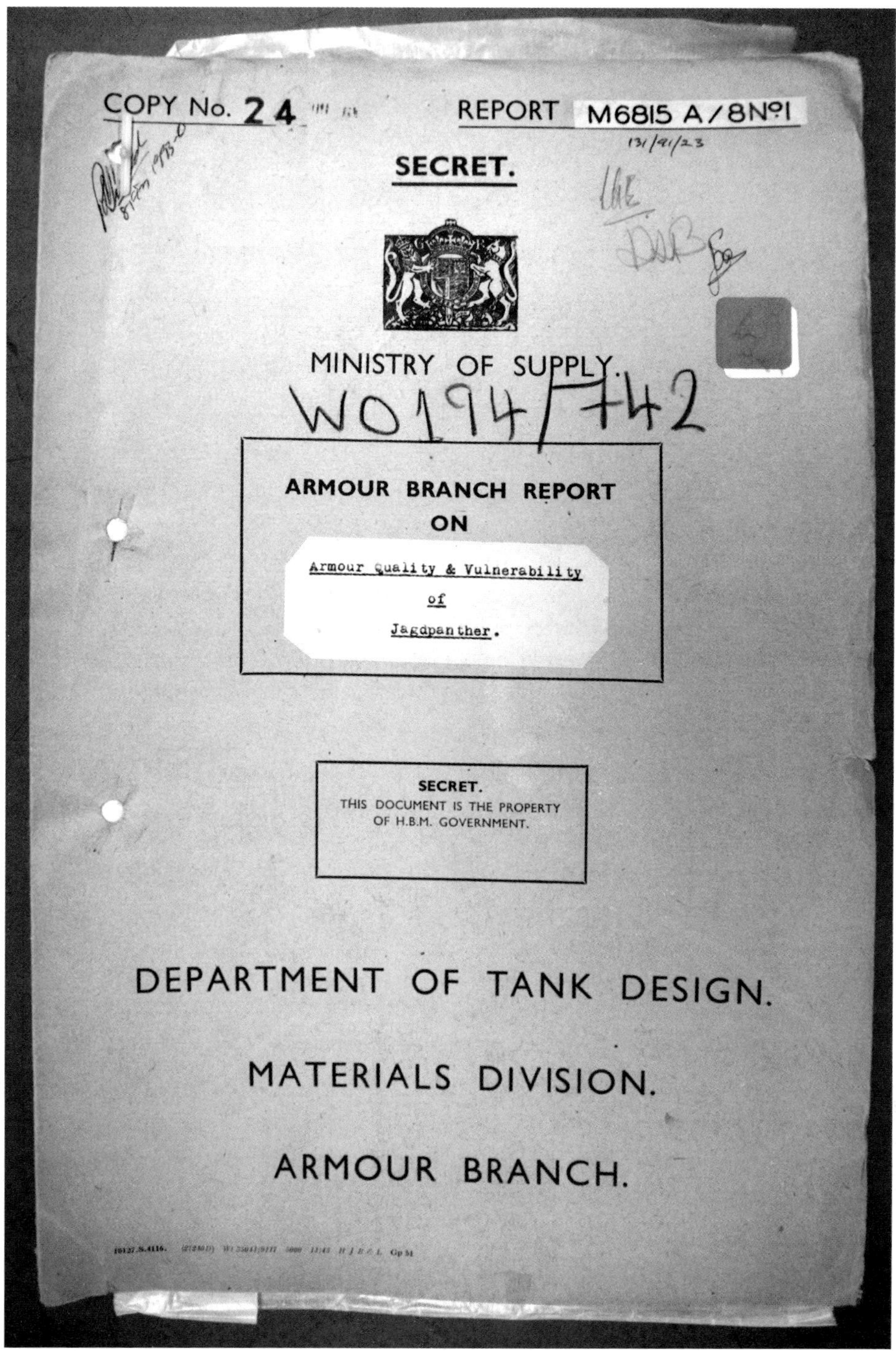

This is what the front cover of a Second World War British War Office report looks like. This particular one is on the armour quality and vulnerability of the Jagdpanther (covered in *The Panzer Archives* Volume 2). Most are held in the National Archives at Kew, London, but copies can be found in archives around the world. This is the front cover of copy number 24.

1

British Testing of the Czechoslovakian-built Praga TNH-P 8-ton Tank (Panzer 38(t)) in 1938

I found an unusual report in the National Archives in Kew that hinted there would have been a remote possibility that the 1939 British Expeditionary Force, sent to defend France, could have been issued with the same Czechoslovakian-built tanks that the Germans equipped their panzer divisions with and used during their May 1940 Blitzkrieg attack of Belgium, Holland

This is the Czechoslovakian-built Praga TNH-P 8-ton tank that was sent to England for testing in June 1939.

and France. The German designation for this tank was the Panzer 38(t). This is the information from that report.

CZECHOSLOVAKIAN-BUILT PRAGA TNH-P 8-TON TANK

On 13 June 1939, the British War Office Mechanization Experimental Establishment (MEE) at Chertsey received a new Czechoslovakian tank by rail, for examination and testing. It was manufactured by Českomoravská Kolben-Daněk (ČKD) who were based near the capital Prague. The vehicle was designated the Praga TNH-P 8-ton tank. It was unpacked by ČKD company's fitters who had accompanied the tank. The tank was completely equipped apart from the ammunition. ČKD were keen to sell their new tank to the British Army and other foreign powers.

This tank was designed to replace the Czechoslovakian Army's LT vz. 35 tank and also be an export success. It had a roomier interior than the earlier tank and a different suspension system. It was armed with a Skoda 37mm gun in the turret and had two 7.92mm Zbrojovka Brno vz.37 machine guns, one in a hull mount and the other mounted in the turret. The armour on the hull front was 25mm thick and 15mm thick on the sides and rear. The armour on the test vehicle sent to England was made of mild steel with the exception of the turret front, which was armoured.

THE DRIVER'S POSITION

The British inspection team first examined the driver's position, noting that it was on the 'offside of the vehicle.' This was unexpected for a European tank as most have the driver's position on the left side of the tank and not on the right. That was the first plus mark noted on the report card. [Editor's Note—Czechoslovakian drivers drove on the left side of the road, just like in Britain, until March 1939 when the commander of German occupation forces ordered a change over to the right side of the road to conform with German traffic legislation.]

The inspectors recorded that the driver's seat was adjustable for length but not for height and that the angle of the backrest could be adjusted. When the front vision hatch was locked in the open position, the driver looked through an opening that was 8 inches (20.3cm) wide by 4 inches (10.16cm) tall. This gave an adequate vision arc of 120 degrees. In wet or dusty weather, a temporary glass windscreen could be locked into position. In combat situations, when the driver's vision hatch was locked in the closed position for protection, an episcope was swung into position. Another means of vision to the front when the hatch was closed was for the driver to peer through the vision slit in the armoured hatch. It was 5 inches (12.7cm) wide by 3/16 of an inch (4.76mm) tall. Bulletproof glass, which was normally stored underneath the driver's legs, could be quickly placed behind the vision slit. A small periscope gave limited vision to the right side of the tank, but the driver could not see to the left. He had to rely on the tank commander and the hull machine gunner sitting on his left to see that everything was clear.

The two steering tillers when drawn back engaged an epicyclic gear by clutch withdrawal and brake application to the plant ring. By pressing a knob on the end of the handle an alternative brake could be applied which operates on the spider, thereby locking the track. Thus, the driver can steer by either epicyclic or clutch and brake methods. Communication between the driver and the commander was by a system of coloured lights.

THE HULL GUNNER'S POSITION

The hull gunner was seated in a similar seat to the driver on the left side of the tank. He was also the radio operator and had to act as the co-driver as previously mentioned. His position was rather cramped due to the wireless sets being placed on a level with his left shoulder. The radio had two alternative aerials, one being a 10-foot vertical rod giving a range of 5kms, and the other being a 'battle' aerial carried on the running board and gave a range of 1km.

The machine-gun on his right was fitted in a ball mounting. It had a limited traverse to the left and right due to the height of the sprocket wheels and mudguards. The gunner's telescope had neither brow pad nor eye padding. This would cause injury if used on the move. The ammunition belt of a hundred rounds was fed into the feed block, the remainder of the belt was suspended on guides from the roof, the whole belt being fed out of the ammunition box.

The machine-gun can be clamped in a central position and fired by the driver who had a remote control on his nearside steering tiller. He sights the gun through an open sight visible through his periscope. The sight was a metal rod about 12 inches tall with a ring on the end. The base of the rod was attached to the glacis plate in front of the driver's position.

TURRET GUNNER

The turret gunner, who was also the tank commander, had a canvas sling seat. He was provided with a non-rotating cupola which had three small periscopes and an episcope mounted on its four sides. He was assisted by a loader who operated the coaxial machine-gun. The examination team commented on the report, 'The optical apparatus, though ingenious, does not give as good vision as the (British) War Department equivalent.' The 37mm main gun and coaxial machine-gun could be fired singularly or both together. The main gun mounting could be either elevated by shoulder control or by a gear control, at will, the firing trigger being on the handle of the latter. When the gun was being fired, the mounting was locked in the position adopted and could not be elevated or depressed so long as the gun was firing. The turret can either be rotated by a hand traversing gear or by free traversing. Locks are provided for both the turret and the gun mountings for travelling. There was no internal turret basket. Ninety rounds of 37mm shells are carried in boxes of 6 rounds. Usually, 30 of these rounds would be armour piercing, and the remaining 60 would be high explosive shells. Each armour-piercing shell weighs approximately 2lb. the high explosive shell weighs 1.8lb. There was stowage for 2,700 machine-gun rounds carried in 100 bullet belts, 3 belts fit in each ammunition box. Nine hundred of the rounds were armour piercing.

THE HULL

The British examiners looked at the tank's fire precautions and the means of exit available to the crew in an emergency. The tank had two main crew hatches, one through the cupola hatch in the turret and another above the hull gunner's head. 'Both are adequate,' was their conclusion. It was noted that the driver did not have his own exit hatch but had to either get out through the turret hatch or if that was blocked, clambered over to the hull gunner's position and get out through his hatch. It was also recorded that it was possible to get into the engine compartment through a small door in the offside internal bulkhead and to open the louvres from inside and get out that way. A large fire extinguisher was conveniently mounted on the wall of the fighting compartment.

THE ENGINE

The tank was powered by a Praga TNHPS/II 4-stroke, 6-cylinder in-line 125hp petrol engine. A hand crank could be used to start the engine from inside the tank as well as from outside the vehicle. A mechanical governor limited the engine speed to 2,000 rpm. The maximum speed of 42km/h (26mph) was based on an engine speed of 2,200 rpm; therefore, the top speed at the governed 2,000 rpm was only 38km/h (23.6mph). The engine was cooled by water circulated by a pump driven off the timing gear. The radiator was mounted at the rear of the engine. Air was drawn in through the louvres under the engine covers, one on each side. It could also be drawn from the fighting compartment by opening slots in the bulkhead. The air inlet to the fighting compartment was by opening an adjustable flap over the brakes and two small louvres. 'It was not considered adequate. Steering gear pollutes the air with hot Ferodo and oil fumes,' the inspectors remarked. Air was drawn through the radiator by a 'Keith' type exhauster and out through a bullet-proof louvre facing upwards on the rear of the tank. This exhauster was coupled to the crankshaft through a universal joint. There was a slipping clutch incorporated in the fan hub. The system did not contain any pressure valves. The vehicle exhaust was very quiet, and on cross-country work, the whole vehicle was extremely quiet compared to other tanks, but it was very noisy on roads due to track noise.

No engine oil cooler was fitted, but the large cylindrical body of the oil cooler was finned and afforded some cooling properties. A large oil bath filter was used for filtering the engine air. A large 'Autoclean' filter was fitted in the lubricating system of the engine. This also incorporated the relief valve for oil pressure. All petrol and oil pipes were of a flexible rubber and canvas hose type, secured by clips. The petrol tanks were in the engine compartment, one on each side. They held 24 gallons each. Petrol was drawn from the tanks by an A.C. engine operated pump. An electric 'Autopulse' pump was also fitted for emergency use.

The transmission was through a single plate clutch in the flywheel. This could not be withdrawn, however, as its only purpose was to give a 'slip' if the Wilson gearbox engaged too fiercely. The power was then transmitted by a propeller shaft through the fighting compartment to a Wilson five-speed and reverse box situated between the hull gunner and driver.

TRIALS

This tank underwent tests from 17–29 March 1939. The weight of the vehicle fully loaded was 9.4 tons. It completed 188 miles (302.5km) by road and 103 miles (165.7km) cross country. The examiners made the following comments:

The commander's field of view was not ideal. The vision from the episcope and the three periscopes was not continuous. It was also extremely hard to judge distance through these instruments. The commander was also hampered when looking through his scopes owing to there being no brow pad. The hull gunner's field of view was adequate to cover the ground over which he could fire. The driver's vision was adequate except for road driving in traffic as the driver needs one member of the crew to be observing on the outside of the vehicle. The driver's position was comfortable except that there was not enough headroom. The hull gunner's position was rendered uncomfortable by the wireless set, causing him to lean continuously to one side. He also suffers from a lack of headroom. The commander's position was satisfactory, with the exception that the sling seat provided did not allow him to adopt a comfortable position behind the gun. The vehicle when closed down does not appear to be adequately ventilated, and fumes given off by the steering gear are very unpleasant after a time.

The power of manoeuvre was adequate and does not vary whether opened up or closed down. The vehicle was also easy to handle on side slopes. The steering requires a little skill as the action of the epicyclic break bands was rapid. Unless the brake was applied skilfully, the tank will turn more than was required when driving on roads. The controls are well-placed. The vehicle does not skid under normal conditions and was safe at any speed it can attain. It does not suffer from reverse steering, but when descending hills the steering becomes very insensitive and heavy. The vehicle was not very large and was inconspicuous as a light tank. The balance of the turret was difficult to estimate as it was extremely awkward to traverse under any circumstances. The traversing handle was very badly placed by British standards. It was to be operated while looking out of the cupola and not while looking through the telescope.

The suspension of the vehicle renders it unsuitable as a gunnery platform. It had a short sharp juddering motion of about two inches pitch which renders it impossible to keep the eye to the telescope. Apart from this, it was quite well sprung and rides across country about similar to the (British) tank, cruiser, A9, Mk. 1. On roads, the suspension was at times affected by a juddering motion, but otherwise, it was satisfactory.

The capacity of negotiating natural obstacles was not adequate for a cruiser tank. It will cross a 5-foot stream but fails to cross a 6-foot stream due to the back falling in as the bank gives way. It will not climb a 4-foot sandbank; the sprocket fails to pull the nose up. It can be fitted with seven spuds on each side of the vehicle's tracks. These spuds are quickly attached to the track, but the short length of the vehicle will not enable it safely to climb more than 3-foot vertical obstacles. It was estimated that the vehicle could cross a 7-foot hard-sided trench. The vehicle climbed a 2 foot 10 inches wooden vertical obstacle. This was the safe maximum owing to the angle to which the vehicle tips itself.

The tank was driven continuously for 94 miles on roads. It took 4 hours 35 minutes, and the average speed was 20.5mph. The average fuel consumption was 3.13mpg. Fuel consumption over cross-country courses was 2.1mpg. After a total of 291 miles, the oil levels did not need topping up. The life of the brakes appeared satisfactory. On a 188-mile journey to Lulworth Ranges, they did not require adjusting. They were only adjusted once after about 260 miles. Two engine stoppages occurred after the vehicle was being tested due to the changing from one fuel tank to the other.

The attempt to produce an inconspicuous machine with observation arrangements immune from bullet attack, has resulted in a cramped fighting machine with controls inferior to our standards. The 'dance' of the vehicle ... is particularly marked on roads and is due to the combination of long pitch narrow bar tread tracks and un-dampened suspension.

CONCLUSION

The British rejected purchasing the Praga TNH-P 8-ton tank because it was inferior to the current British Cruiser tanks in its ability to cross obstacles, lack of smooth ride, poor ventilation and cramped fighting compartment. It was too thinly armoured to be considered an infantry tank. Its Skoda 37mm gun was not as powerful as the British 2pdr gun. The tank was returned.

As an interesting side note, in 1934 Skoda submitted a patent for a track pin retention incline guide (a track pin bump). Each track pin has a head at the end nearer the hull. A track pin looks like a very big nail. There is nothing at the other end to prevent it from coming out, but on the hull is fixed an inclined guide or cam wiper plate. If the pin starts to come out of the track link its head engages this plate as it passes it and the pin gets pushed back into place. These track pin bumps were used by the Soviets on their T-34 tanks. The Germans also used them on their Panther and Tiger tanks.

The Czechoslovakian-built Praga TNH-P 8-ton tank driver could only see forward and to the right.

The Czechoslovakian-built Praga TNH-P 8-ton tank was armed with two 7.92mm Zbrojovka Brno vz.37 machine guns and a Skoda 37mm gun.

The Czechoslovakian-built Praga TNH-P 8-ton tank track was powered by the sprocket wheel at the front of the tank.

Two Panzer 38(t) tanks on display at the Vadim Zadorozhny Museum, Arhangelskoe, near central Moscow. (*Yuri Pasholok*)

A Panzer 38(t) tank on display at the Patriot Park Museum, Kubinka, Russia. (*Yuri Pasholok*)

Czechoslovakian LTH export tank (Panzer 39 in Swiss Army Service) in a running condition kept at the Schweizerisches Militarmuseum in Full, Switzerland. (*Massimo Foti*)

Czechoslovakian LTH export tank (Panzer 39 in Swiss Army Service) in a running condition kept at the Schweizerisches Militarmuseum in Full, Switzerland. (*Massimo Foti*)

Czechoslovakian LTH export tank (Panzer 39 in Swiss Army Service) in a running condition kept at the Sammlung Historische Panzer in Thun, Switzerland. (*Massimo Foti*)

Czechoslovakian LTH export tank (Panzer 39 in Swiss Army Service) on display at the Sammlung Historische Panzer in Thun, Switzerland. (*Massimo Foti*)

Licensed-built Panzer 38 S based on the Czechoslovakian LTH export tank on display at the German Tank Museum. (*Massimo Foti*)

2

The Panzerkampfwagen II Ausf.A Light Tank

In January 1934, the German tank design office of the weapons testing ordnance department Waffen Prüfwesen 6 (Wa Prw 6) issued specifications of a new tank hull they wanted built, code name La.S.100. Weapons manufacture Maschinenfabrik Augsburg Nürnberg AG, (M.A.N.) constructed a prototype La.S.100 tank hull. They competed with two other German companies, Fried. Krupp Abt.A.K. and Henschel. M.A.N. was awarded the contract to build the

Panzer II Ausf.A tank three-quarter view showing additional armour plating on hull front and front vertical plate, also on mantlet and glacis plate. There are curved plates on the mantlet and angled plates on the turret sides. The guns have been removed.

hull of the new Panzer II light tank based on their prototype La.S.100 hull. Daimler-Benz designed the superstructure and turret.

It is wrong to dismiss the Panzer II tank of 1936 as a poor design when comparing it with later manufactured, more heavily armed and armoured tanks of the Second World War. The tank's armour could protect its crew from small arms fire and 7.92mm S.M.K steel-cored armour-piercing machine gun bullets fired from a range of 30m. It was designed to attack enemy machine gun nests and destroy them so that the accompanying infantry could continue to advance. The Panzer II was not primarily designed to engage in tank-on-tank combat. The tank's 2cm Kw.K.30 L/55 gun could knock out Soviet T-26 and BT tanks, but the crews were aware that the Panzer II tank's armour would not stop a 3.7cm or 4.5cm anti-tank gun. The high nickel-alloy, rolled homogeneous-hard armour plate ranged in thickness from 5mm to 13mm. It was welded together not riveted as seen on many other tanks of this time period. This made it stronger and lighter.

The first Panzer II Ausführung (model versions) were given the lower-case letter 'a' then 'b' and 'c'. Later versions were given capital letters 'A', 'B' and 'C'. This can be confusing. The Panzer II Ausf.a tanks were subdivided into Ausf.a/1, Ausf.a/2 and Ausf.a/3. Each version had different minor mechanical changes. Early versions of the Panzer II changed shape over time as they were upgraded during their operational life. Additional armour was added, and features like cupolas were fitted. Panzer II tanks were not used in the Spanish Civil War. They first saw combat in Poland, 1 September 1939.

The Panzer II Ausf.a and Ausf.b were powered by a Maybach HL 57 TR 6-cylinder water cooled 130hp petrol engine that gave the tanks a maximum road speed of 40km/h (35mph). They had an operational range of 190km (118 miles). In addition to being armed with a 2cm Kw.K.30 L/55 auto-cannon, a 7.92mm MG34 machine gun was mounted in the turret next to the main gun. The 2cm Kw.K.30 gun could fire three different shells. When fired against armour plate laid back at 30 degrees from the vertical, the PzGr.39 (Armour Piercing) shell could penetrate 23mm of armour at 100 metres and 14mm of armour at 500 metres. The PzGr.40 (Armour Piercing Composite Rigid) shell could go through 40mm of armour at 100 metres and 20mm of armour at 500 metres. It could also fire 2cm Sprgr. 39 (High Explosive) shells.

The thickness of armour on the Panzer II Ausf.b tank's hull, superstructure and turret were increased from the Ausf.a tank's 13mm to 14.5mm. The gun mantlet's armour increased from 15mm to 16mm. To reduce the reliance on obtaining nickel, the armour was changed to rolled homogeneous nickel-free steel armour. It had the same resistance to 7.92mm S.M.K steel-cored armour-piercing machine gun bullets fired from a range of 30m as the Ausf.a, but it had to be thicker to achieve this. This increased the weight of the tank by 500kg, but it did not decrease its speed.

The shape and thickness of the crew's vision ports were changed to give added protection. A different style of large drive wheel was bolted on to the final drive at the front of the tank. The rear engine deck was redesigned. Armoured louvres were added to the rear right of the tank. The road wheels and the track return rollers were widened. The return rollers were reduced in diameter. Wider tracks, increasing from 260mm to 285mm, were introduced. Lengthened, foldable track guards were fitted to the rear of the tank.

The track on the Panzer Ausf.b and Ausf.a was driven by a large sprocket wheel at the front of the tank. The idler wheel was at the rear. There were six small road wheels attached to three bogies on a metal bar. At the top there were three return rollers. This changed on the Ausf.c. The suspension on the Ausf.c was visually very different from that used on previous models. Five larger 55cm diameter road wheels replaced the six small road wheels. The suspension was now a leaf spring, crank arm system. The long metal beam that ran along the road wheels was no longer needed and was removed. The new version of the front-drive wheel first introduced on the Ausf.b was kept. An additional track return roller was added, bringing the total to four. The front track guard extension was now held together with clips. This increased the total weight from 7.9 tonnes to 8.9 tonnes. This did not affect the tank's top speed, as the engine was upgraded as well. It was fitted with a more powerful Maybach HL 62 TR 6-cylinder water-cooled 140hp petrol engine. Turret ring guards were not present on this vehicle, but some had them back fitted. The additional armour added to the front hull glacis plates changed the look from a curved frontal armoured hull to an angular shape. The factories produced 75 Panzer II Ausf.a, 100 Panzer II Ausf.b, 75 Panzer II Ausf.c, and 210 Panzer II Ausf.A.

The Panzer II Ausf.A was the final standardised version ready for mass production. The previous versions Ausf.a/1, a/2, a/3, b and c were all trial series developed to test new design elements. This is why a capital letter 'A' was used to denote the production version. Only minor internal changes were made. A new gearbox was fitted. The fuel pump, oil filter and cooler were relocated on the engine. The tank's electrical system was suppressed to try and stop it interfering with the AM radio reception and transmission.

The main visual difference between the Ausf.c and the Ausf.A was the introduction of a new driver's visor at the front of the tank. The large flat rectangular armoured vision port cover was now replaced with a V-shaped armoured visor that had a slit built into it. The two side visors used by the driver and radio operator were now of the same type. The early versions of the Panzer II Ausf.A did not have a turret ring guard bolted on to the superstructure at the front and rear of the turret ring.

THE CAPTURED PANZER II TRIALS

A Panzerkampfwagen II Ausf.A light tank was captured by the Allies in Libya and shipped back to the Fighting Vehicle Proving Establishment in Chertsey to obtain information on its reliability and performance. It had a few modifications fitted, such as the stowage bin on the right track guard, bolted-on armour, a cupola with periscopes and larger louvres in the radio operator's hatch to keep it cool in the desert heat. It had a white bisected letter 'D' painted on the front superstructure armour plate to the right of the driver's vision port and at the rear of the tank. This was the insignia of the 21st Panzer Division. This Panzer II Ausf.A did not end up in the Tank Museum at Bovington. Their Panzer II is an Ausf.F. The fate of this Panzer II Ausf.A is not known. There are only nine surviving Panzer II tanks left in the world and this tank is not one of them. The assumption is that it was sold for scrap metal. What follows is the transcript of the British trials report on that captured Panzer II Ausf.A tank. The abrupt military terminology and reporting style has been kept. The British used the abbreviation Pz.Kw to stand for the German term Panzerkampfwagen (armoured combat vehicle). The German abbreviation is normally Pz.Kpfw.

SECRET

FIGHTING VEHICLE PROVING ESTABLISHMENT FIELD TRIALS REPORT ON GERMAN LIGHT TANK PZ.KW.II (EX-LIBYA)

F.V.P.E. FIELD TRIAL REPORT NO. F.T.679

STATEMENT

It was required to carry out road and cross-country performance trials with the Pz.Kw.II in accordance with standard practice. A general report on this machine together with photographs is included. No parts of the machine were dismantled. Due to certain mechanical faults, Project No.E.1293 on Pz.Kw.II (ex-Russia) was transferred to this project No.E.1296 Pz.Kw.II (ex-Libya)

CONDITIONS

Weight of Vehicle: Unladen 8 tons 19 cwt. Laden 9 tons 19 cwt.

Note: As the weight of equipment carried is unknown, the machine was laden to correspond with that carried by the equivalent British type, the Tetrarch.

Mileage of Vehicle on dial: 305 miles + 103cc total 408.

Total speedo reading at the conclusion of the trial 7797.5 miles

DIMENSIONS

Overall length	15' 5"
Overall width	7' 4"
Overall height	6' 5 ¾"
Belly clearance	1' 1"
Track centres	6' 2"
Length of track on ground on concrete	9' 2"

HULL ARMOUR

Front	15mm + 20mm additional plates
Side	15mm
Back	15mm
Top	10mm

TURRET ARMOUR

Front	15mm + 20mm additional plates
Side	15mm
Back	15mm
Top	10mm
Mantlet	15mm + 15mm additional plates

ARMAMENT

One 2cm (Kwk.30) automatic cannon and on M.G.34 (7.90mm) co-axially mounted, sighted by telescope and open sight. The trigger for the 2cm is incorporated in the elevating handwheel. The trigger for the M.G.34 is incorporated in the traversing handwheel. Some models may be fitted with the 2cm tank gun (Kwk.38)

Traverse—360° hand traverse—elevation by handwheel

Ammunition—2cm—18 magazines each holding 10 rounds of AP, AP incendiary or HE. M.G.34 1600 round.

ENGINE

Make—Maybach O.H.V. Model H.2. 62 T.R.
Type—Straight six-cylinder, four stroke petrol
Bore—105mm
Stroke—120mm
Capacity—6.191cc
Valves—O.H.C. and rocker
Rating—@260rpm 135hp

[The above data extracted from M.I.10 Intelligence summary No.73 dated 21 May 1942]

OBSERVATIONS

The engine is mounted at four points, and located in a compartment at the offside rear of the tank, and is accessible through two cover doors in the top plate: access can also be gained by removing portions of the engine bulkhead inside the machine, and front radiator compartment to the nearside of the engine. A flexible pipe in the offside carries the exhaust gases to the silencer mounted transversely at the rear of the machine. Two smaller pipes presumably outlet and inlet are taken from the exhaust manifold to an annulus surrounding the mixing chamber of the carburettor. Also located on the offside of the engine is the electric starter motor, which is solenoid operated.

On the nearside of the engine will be found from front to rear, the inertia starter, petrol pump (Pallas type) driven from the crankshaft and fitted with hand primer and bowl filter, and an impellor type water pump which is driven in tandem with the dynamo. The dynamo, which is driven by V-belt from a pulley on the crankshaft extension, is of the ventilated type drawing cool air by way of a flexible duct from the fighting chamber. The carburettor and induction manifold are located directly above the water pump. Driven by V-belt from the crankshaft extension, and at the nearside rear of the engine compartment, is an enclosed fan which draws cool air by way of ducts over the steering breaks.

METHOD OF STARTING

1. 12v electric starter (solenoid operated).
2. Inertia starter, wound from the rear and operated either from the rear outside the machine, or from inside the fighting chamber.
3. By starting handle at the rear of the machine.

IGNITION SYSTEM

By Magneto (Bosch impulse starter type SR.6 R12P) driven from the timing gears; the six 14mm sparking plugs (Bosch W. 145122) are recessed behind the cover plate in the nearside of the cylinder head. The magneto and HT leads are screened against wireless interference.

COOLING SYSTEM

Close system with a capacity of 7 gallons 4 pints. A pressure valve and filler are located in the header tank over the front of the engine. An impeller type of water pump is driven in tandem with the dynamo, and twin V-belts, driven from pulleys on the crankshaft extension, drive a 10 bladed cast aluminium fan mounted at the rear of the radiator. The radiator is of the fin type and mounted to the nearside of the engine. The hinged metal blanking plate is accessible from the fighting chamber. The airflow is through louvres in the offside of the hull and over the cylinder head, through a grille in the top plate, and from the fighting chamber. Hot air is then exhausted through the grills at the rear of the engine and radiator cover plates.

PETROL SYSTEM AND CARBURATION

A single four float Solex down draught carburettor (40 JFF2) of the duplex type, is fitted to the right side of the engine. Starting from cold is obtained by the normal Solex starter carburettor fitment. The mechanical type of pump driven from the crankshaft and fitted with filter bowl and hand primer is located on the right of the engine. Two petrol tanks fitted in the offside of the fighting compartment have capacities for 16 gallons 3 pints, and 22 gallons 2 pints, respectively; access to the fillers is obtained by removing circular cover plates in the offside of the top plate of the fighting chamber.

LUBRICATION SYSTEM

Dry sump, four-speed. The system has a capacity for 2 gallons 6 pints. The oil tank is located in the offside rear of the engine compartment, the dipstick being incorporated in the filler cap. An oil cleaner of the Auto-Klean type is operated manually by remote control from the front engine bulkhead.

TRANSMISSION

The propeller shaft, running close to the petrol tank, and enclosed by a metal guard, transfers the drive from the flywheel to the clutch, which is incorporated with the gearbox. A rubber disc type of universal coupling is fitted to the front and rear of the propeller shaft.

GEARBOX

The gearbox was of the crash type marked SSG-46-ZF Getriebeis mounted to the right of the driver and has six forward speeds and one reverse. The full correct designation was SSG-46-ZF, synchronisiertes Schaltgetriebe 46 Zahnradfabrik Friedrichshafen (synchronized manual transmission model 46, gears manufactured in Friedrichshafen). Lubrication is by pressure, and a manually operated Auto-Klean type of cleaner is mounted on top of the gearbox. A small foot operated catch allows the engagement of first and reverse gears. Oil filler and breather are integral with the gearbox casing.

BEVEL BOX

Forward of the gearbox, the bevel box transfers the drive at right angles to the steering units. The speedometer drive is taken from the bevel box, and at the front of the box is a double worm type of oil pump which supplies the required lubrication for the steering units.

STEERING UNITS

The steering units are of the epicyclic type. The first movement of the steering lever releases the external contracting cast iron brake on the annulus of the steering epicyclic, further movement of the steering lever applies a skid brake which is mounted in tandem with the epicyclic, and presumably conveys the drive to the sprocket from the spider of the epicyclic.

FINAL DRIVE

It was not possible to examine the final drive, but there appears to be still further reduction in the outside casing before the drive finally reaches the sprocket.

TRACKS AND SUSPENSION

The tracks are similar to those fitted to the old Carden-Loyd tractors. The links are of the single pin type, positioned by S hooks. Track measurements: width 11 3/16-inch, pitch 3 5/8-inch, diameter pin 5/8 inch.

Five 22" diameter disc type bogie wheels on each side, equally spaced, are constructed from light alloy, with rubber tyres, width 3.5" and depth 2.75". Tyre markings read 550/100–455 Continental. The wheels are mounted on 'L' shaped arms which are pivoted at an angle, to the hull. To the top of the arm is secured a quarter elliptic spring; the floating end of the spring is free and bearing on rollers fixed in the hull. A rubber faced stop is fitted above the suspension arm to limit the upward movement of the wheel. The rigidity of the suspension causes the vehicle to ride badly on cross-country drives. Spring breakages are frequent. The sprocket is 2' 7" in diameter and has 26 teeth. There are four rubber tyred top rollers in each side, of 8 ½" diameter. Steel rimmed idler wheels, diameter 2' 1 ¼" are mounted on brackets at the rear of the machine. Track adjustment, which is an extremely simple operation, is affected by removing a serrated keeper plate on the bracket casing, and by the use of a special tool, rotating the idler wheel spindle by means of a hexagon extension at the end of the spindle.

DRIVER'S CONTROL

The driving position is quite normal, the driver sitting slightly to the nearside in front of the machine. The clutch pedal is on the left, accelerator centre, and foot brake on the right. The steering levers are easy in operation and are fitted with spring loaded ratchets for parking purposes. A small foot operated catch is located at the side of the gearbox to enable the gear change lever to be brought into line with first and reverse gears.

INSTRUMENTS

The instrument panel to the driver's right contains the ignition switch, starter switch, and head and tail lamps indicators, a large revolution counter calibrated up to 3,500rpm, and a speedometer. A further panel on the offside hull plate contains the oil pressure gauge, and thermometer. Directly behind the gearbox, is the main junction and fuse box.

FIGHTING CHAMBER

The wireless receiver and transmitter are located in a recess in the rear of the hull. The HT supply unit for the set is to be found under the floorboards at the rear of the fighting chamber. The aerial can be moved from the working position to the rest position by means of a lever on the left of the wireless operator.

The Commander/Gunner's seat rotates with the turret. There is an electrical base junction, which supplies current to all rotating parts. The crew's positions are rather cramped, the diameter of the cupola being 17 ½". Entry and exit are not easy for a large man. The cupola is non-revolving, but eight episcopes are fitted giving ample all-round vision. Four observation ports are located in the sides and rear of the turret.

HULL

The hull is welded and fabricated except where additional 20mm plates are fitted in the following positions:

  I. Turret, front.
 II. Front vertical plate.
III. Front glacis plate.
IV. Front nose plate.

ORIGIN

This trial was initiate by V/GD to obtain information on the reliability and performance of a German Tank Pz.Kw.II ex Libya.

REMARKS

The machine was not a new one and the high speedometer reading before commencement of the trial should be taken into consideration. The machine behaved extremely well and little serious mechanical trouble was experienced.

SIGNED

A.T. Sweeney Assistant Director (General Design) Fighting Vehicle Proving Establishment.
8 March 1943

Panzer II tank with regular spaced bogies suspension. The wireless aerial cradle can be seen on the left side of the tank on top of the track guard.

This Panzer II tank driver's escape hatch was in front of his vision ports. The armour on the turret side, top and mantlet was curved. A bullet splash turret ring guard can be seen in front of the turret.

The engine of the Panzer II was at the rear of the tank. Armoured louvres were fitted to keep the engine cool to enable air to be sucked in and hot air vented out. The black arrows indicate the direction of the air.

Panzer II engine compartment. The jacking block, starting handle extension, S hook and crowbar were stowed on the track guard.

Photograph of the Panzer II engine compartment from the left side. On the front is the driver's escape hatch, behind that was the gearbox and clutch housing. The upright tubes are the petrol filters.

The Commander's cupola in the tanks' turret is non-revolving, but eight episcopes are fitted giving ample all-round vision.

The protruding bolts shown in the circles were added by the British. They were used to attach additional weight bars to simulate the combat weight of the tank fully loaded and to compensate for the missing gun and crew members.

The Panzer II tank turret ring.

3

Poison Gas Bullet Attack Trials on a Panzer III

During the Second World War, if the Nazis started to use chemical weapons, the Allies were ready to respond in kind. Scientists at the Chemical Defence Experimental Station (CDES) in Porton Down, experimented with poison gas sprays, aerial gas bombs, concentrated pepper spray tipped bullets and pepper spray grenades that were designed to be fired at advancing enemy tanks. They needed to establish how effective they would be as they relied upon the tank's ventilation system to suck the gases into the fighting compartment. The gas used was Hydrogen Cyanide (also known as Prussic Acid) which is a chemical compound with the formula HCN and structural formula H–C≡N. It is colourless, extremely poisonous, and flammable. The word Lachrymatory (causing tears) is used in the report. This refers to CAP weapons that use

Attack on a Panzer III tank with the Squirt Anti-tank portable Mk.1 charged hydrogen cyanide.

Capsaicin. This is a non-lethal compound that causes temporary incapacitation and a burning sensation in the eyes and throat, very much like concentrated pepper spray. This is the trials report.

SECRET

EXPERIMENTS ON THE RATE OF CLEARANCE OF GAS FROM TANKS PART V—GERMAN PZ.KW III

22 MARCH 1943

The efficiency of the ventilating system of the Pz.Kw III tank in dissipating concentrations of gas from the interior has been assessed by the method described in Priority Program No.2555 Part I, Churchill Tank. The results indicate that at an engine speed of 2,000 rpm the effective ventilation rate of the turret is about 4.5 changes of air per minute, and that of the forward part of the tank only 1.7 changes per minute.

Comparison with the results of previous trials shows that the Pz.Kw III tank conforms closely from the ventilation point of view to the average of several British and American designs. It is also apparent from this comparison that tanks such as the Pz.Kw III, which are ventilated by utilizing the engine induction air, are inferior as regards the clearance of gas to the majority of those in which the engine cooling fan produces the ventilation of the interior.

INTRODUCTION—VULNERABILITY OF TANKS TO ATTACK WITH CHEMICAL WEAPONS

1. The experiments dealt with in this report were made in order to assess the efficiency of the ventilating system of the Pz.Kw III tank in dissipating concentrations of gas inside the tank, and to obtain a comparison with British and American types.

THE VENTILATING SYSTEM OF THE PZ.KW III

2. The ventilation system of the German Medium tank Pz.Kw III is illustrated by a drawing attached to this report.
3. This system is similar in its main features to that of the Infantry tank Mk.IV (Churchill). The main ventilation is affected by the induction system of the engine. The carburettor intake pipes enter the fighting compartment of the tank through a bulkhead plate on which air filters are fitted. The air thus drawn from the fighting compartment is replaced by outside air which is sucked through cowls, gun mantlets and other minor openings in the structure of the hull and turret. There is also an auxiliary extractor fan in the roof of the turret

to deal mainly with gun fumes. As this fan had been removed from the tank before its arrival at Porton Down, a fan from a Churchill tank was substituted.

4. A feature, not present in the Churchill tank, was a fan driven by the engine drawing air through the forward brake drum casings which were situated far forward in the interior of the tank. The inlets to the casings were close to small cowls on the front of the tank so that the air was drawn almost directly into the drums and hence had little effect on the ventilation of the interior compartments.

5. The volume of air in the interior of the tank (allowance being made for the volume of the crew) was approximately 5.7 cubic metres (200 cubic feet).

6. The interior of this tank was not divided into separate compartments (turret, driving compartments etc.) but was one chamber with little in the way of obstructions.

METHOD OF EXPERIMENTS

7. The standard procedure adopted in the previous experiments with British and American tanks for estimating the rate of decay of gas concentration due to ventilation was adhered to. Ampoules containing 100cc of sulphur dioxide were broken in the following positions:

    (i) Just below the breech of the 50mm gun in the turret.

    (ii) Near the driver's legs.

    Eight instantaneous samples were taken at each of four of the crew positions (five in some cases) over a period of three minutes.

8. The tank was stationary and facing into the wind during the majority of the experiments but it was found possible, despite the bad mechanical conditions of the engine and transmission system, to make four experiments with the tank moving in a low gear at about 3mph. The engine revolutions were about 2,000 rpm. Taking into account the throttle opening required to give these engine revolutions under such conditions, it is probable that the consumption of induction air was nearly as great as would have been the case had the tank been in good condition travelling at cruising speed over level country.

9. The proper conditions of the rear bulkheads and engine induction system was not known when the tank was first available, due to the absence of the induction trunking, air filters and the cover plate over a large aperture in the rear bulkhead. Some experiments were done therefore with this aperture in the rear bulkhead.

    (i) completely open.

    (ii) completely closed.

    In each of these conditions the engine induction air was drawn from the engine compartment. With the aperture open, the engine cooling fan was causing air to pass from the fighting compartment through to the engine compartment.

With this aperture closed the fighting compartment had no forced ventilation with the exception of the auxiliary fan and possibly a very small ventilation due to the brake cooling air mentioned above. The cooling arrangement was with the bulkhead aperture closed with a cover plate through which the engine induction trunk passed, the inlet being inside the fighting compartment.

10. The data derived from these experiments are tabulated in the Appendix. The definitions of the various terms used are as follows:

  (i) 'U' is the concentration which would result from the instantaneous distribution of the contents of the ampoule throughout the whole volume of air in the fighting compartment of the tank.

  (ii) The Persistence Time of a concentration is the time during which the concentration exceeds 50% of the maximum concentration.

  (iii) The Decay Co-efficient (B) is derived from the rate of decay of the concentration when it is exponential or approximately so and is numerically equal to the effective rate of ventilation in changes of air per minute.

DISCUSSION OF RESULTS—RATE OF DECAY OF GAS CONCENTRATIONS

11. The rate of decay of the sulphur dioxide concentrations was more regular and more nearly exponential than was the case in the majority of tanks tested previously. This can probably be attributed to the greater simplicity of the interior of this tank, the interior being one chamber without subdivisions.

12. The values of the decay coefficients obtained are tabulated below. The value marked with an asterisk appears anomalously high but is the result of one experiment only.

Table I: Decay Coefficients (B)—changes of air per minute in Pz.Kw. III

Speed	rpm	Conditions of rear bulkhead	Auxiliary Fan	B for turret	B for forward part	No. of Experiments
Stationary	2000	Correct	On	3.9	1.7	4
3 mph	2000	Correct	Off	3.0	–	1
3 mph	2000	Correct	On	4.8	1.6	3
Stationary	1200	Aperture open	Off	2.0	1.2	2
Stationary	2000	Aperture open	Off	3.6	1.3	2
Stationary	2000	Aperture open	On	3.1	1.6	4
Stationary	2000	Aperture closed	Off	1.0	1.3	2
Stationary	2000	Aperture closed	On	3.6*	1.0	2

**Table 1: Decay Coefficients (B)—changes of air per minute in
        Tank Mk.IV Churchill**

Speed	rpm	Conditions of rear bulkhead	Auxiliary Fan	B for turret	B for forward part	No. of Experiments
Stationary	1800	Correct	On	3.0	0.6	5
Stationary	1800	Correct	Off	1.0	0.1	4
Stationary	1350	Correct	Off	0.9	0.1	2
Cruising Speed	1100	Correct	Off	4.3	0.1	4

13. It would appear that when the tank was moving (the greater the throttle opening resulting in increased withdrawal of air from the fighting compartment) the effective ventilation of the turret was increased from about 4 changes per minute to nearly 5. The ventilation of the forward part of the tank was however not affected and varied little under the different conditions investigated.

14. The differences in ventilation between the forward part of the tank and the turret position was well marked and, in this respect, the Pz.Kw III resembled those tanks already tested which had a similar ventilation system where air enters into and is withdrawn mainly from the rear part of the tank.

15. The results of the experiments with the bulkhead aperture either fully open or closed are interesting in as much as they show that the ventilation of the tank could not be improved by introducing an aperture in the bulkhead with a view to utilizing the effect of the engine cooling fan.

16. The effect of the auxiliary fan was to increase the ventilation, but hardly to a significant degree. The forward positions were however about three times better ventilated than the corresponding positions in the Churchill. This was doubtless in part a consequence of the less obstructed interior of the Pz.Kw III but may have been enhanced by air currents induced by the braking cooling system. The forward positions of the Pz.Kw III cannot be said, however, to be well ventilated as the ventilation rate of 1.5 changes per minute is less than half that of the rear positions.

MAGNITUDE OF CONCENTRATIONS RECORDED

17. The maximum concentrations recorded with the bulkhead inserted correctly varied from about 15 to 35% of U, except one fleeting concentration of about 130%, with the SO_2 released in the rear part of the interior and from about 10 to 60% of U when the gas was released forward. The averages were respectively 22% and 35%.

18. The persistence time in the former case averaged about 0.25 mins; in the latter it was greater but variable, probably averaging about 0.5 mins.

MEAN CONCENTRATIONS OVER HALF A MINUTE—GAS RELEASED IN THE TURRET PART OF THE TANK

19. The average mean concentration over the half minute of highest concentrations was 15% of U for the forward positions and, with one exception, only half as great for the turret positions. The use of the auxiliary fan had little effect on these values.

GAS RELEASED NEAR THE DRIVER'S POSITION

20. The forward positions were again subject to the higher concentrations; the concentrations measured in the turret positions were practically negligible. The average 'half minute' concentration for the forward positions was 21% of U and again appeared to be unaffected by the auxiliary fan.

THE TIME TAKEN FOR THE MAXIMUM CONCENTRATION TO FALL TO A LOW VALUE

21. The time after which the highest remaining concentration at any time in the tank does not exceed 5% of U provides an indication of the net effect of two factors:
    (i) The magnitude of the initial concentrations.
    (ii) The rate of decay.

The comparison between the Pz.Kw III and the Churchill Tank is illuminated by the following table II:

Tank	Time (mins)	
	Gas released in turret.	Gas released forward.
Pz.Kw III (fan on 2000rpm)	0.8 mins	1.5 mins
Churchill (fan on 1800rpm)	1.2 mins	5.0 mins

This illustrates the fact that gas released in the relatively stagnant air in the forward part of the Churchill Tank was slow to disperse. The Pz.Kw III is seen to be very much better in this respect.

CONCENTRATIONS DUE TO 100G OF VOLATILE LIQUID

22. The average concentrations resulting from the release of 100g of volatile liquid in the tank can be estimated from these results to be of the order of the values tabulated below.

Table III: Concentrations likely to be established in the German tank Pz.Kw III due to the instantaneous release of 100g of highly volatile liquid

Gas released in	Average concentration during the half minute of highest concentration. g/m^3		Condition of Tank –Tank stationary; engine 2,000rpm or Tank moving at cruising speed 1,500rpm. Auxiliary fan either on or off. $U = 17.5g/m^3$
	Turret	Forward Part.	
Turret Part.	1.75	2.5	
Forward Part.	0.2	3.75	

APPLICATION OF RESULTS TO HCN

23. Taking as a basis the lethal dose of HCN to be $5g/m^3$ for 30 secs, it will be seen that the production of a lethal concentration at least some point in this tank would require the introduction of amounts of HCN of the order of:
    (i) Into the turret part of the tank ... 200g
    (ii) Into the forward part of the tank ... 150g
24. The amounts of HCN estimated to be required to produce a lethal concentration at some point in this and other tanks tested are tabulated below.

Table IV: Effective ventilation rates and amounts of HCN required to produce lethal concentrations in various tanks

Tank	Condition	Gas released in turret		Gas released forward	
		Turret ventilation air changes/ min	Amount of HCN to produce lethal concentration	Ventilation of forward part. changes/ min	Amount of HCN to produce lethal concentration
Churchill Mk.IV	Stationary 1,800rpm. Fan on	3	$150g/m^3$	0.6	$60g/m^3$
Covenanter Cruiser Mk.V	Stationary 1,500rpm.	5	$350g/m^3$	5	$350g/m^3$
Crusader Cruiser Mk.VII	Stationary 1,200rpm.	6	$400g/m^3$	1.7	$100g/m^3$
M3 Grant and Lee	Stationary 1,500rpm.	2-2.5	$70g/m^3$	*	$150g/m^3$

Tank	Condition	Gas released in turret		Gas released forward	
		Turret ventilation air changes/ min	Amount of HCN to produce lethal concentration	Ventilation of forward part. changes/ min	Amount of HCN to produce lethal concentration
M4 Sherman	Stationary 1,500rpm.	6.5	400g/m³	3.0	150g/m³
Allied Average		4.5	270g/m³	2.6	160g/m³
Pz.Kw III	Stationary 1,800rpm. Fan on	4.5	200g/m³	1.7	150g/m³

* = Erratic decay of SO2 concentrations made it impossible to ascertain the effective ventilation rate.

25. It is apparent that the Pz.Kw III tank conforms closely from the ventilation point of view to an average of several British and American designs. In most cases the tanks were more vulnerable to gas released in the forward part of the tank and experiments with the Jet anti-tank charged HCN have shown that the forward positions do in fact experience the higher concentrations. It follows therefore that figures in the right-hand column of Table IV are the more significant (except in the case of the M3 Grant). An examination of this column shows that the Pz.Kw III is better than the Churchill and the M3 Grant and Lee tanks, comparable to the Crusader and M4 Sherman but considerably inferior to the Covenanter—a tank of uniformly good ventilation through the interior.

26. The results of Table IV have been plotted in the form of a graph relating the amount of HCN required to produced lethal concentration in a tank compartment, to the ventilation in that compartment. It is found possible with a quite reasonable degree of accuracy considering the nature of the experiments, to represent these results by a single curve.

27. This graph will enable an estimate to be made of the concentrations likely to be produced by the instantaneous release of gas in any tank for which data of the ventilation rates in the various compartments is available.

28. An examination of the results achieved with the Jet Anti-tank Mk I (formerly known as Squirt A/T) charged HCN against several British and American tanks in the above table (7 experiments in all) showed that, on the average, concentrations over half a minute of the order of 15g/m³ were produced in the driving position, the corresponding figure for the turret being about half as great. The results for the Pz.Kw III conformed closely to this average, the

average values from three experiments being: driving position $13g/m^3$ and turret $8.5g/m^3$. These concentrations indicate that an amount of HGN of the order of 350g entered the driving compartments of the tank in most cases. A consideration of the table leads to the conclusion that the German tank would need a ventilation rate of 10 air charges per minute (a rate difficult to achieve and uncomfortable for the crew) in order to reduce the 'half-minute' concentrations to about $2 g/m^3$—a dose which would not be lethal but might produce unconsciousness in some cases. In default of this, the only alternative method open to the Germans of rendering the Pz.Kw III invulnerable to the Jet A/T charged HCN would be that of reducing the area of apertures permitting the ingress of liquid to about 10%–20% of the existing areas.

CONCLUSIONS

29. The effective ventilation of the Pz.Kw III tank at 2000rpm is about 4.5 air changes per minute in the turret portion and 1.7 changes per minute forward, conforming closely to the average of five British and American types. It is better ventilated than the Churchill Mk.IV tank and the M3 Grant and Lee but worse than the M4 Sherman and the Covenanter Cruiser Tank Mk.V.

30. The amount of HCN required to be released in the turret of this tank to produce a lethal concentration for at least 30 seconds at one or more crew positions, is estimated to be about 200g. The corresponding figure for the forward part of the tank is 150g.

31. The results with this tank and with the Churchill Mk.IV tank indicate that the ventilation produced by utilising the engine induction air (even when supplemented by a small electric fan), is not as great as that achieved in most tanks in which the engine cooling fan produces the ventilation of the interior.

32. From the results with this and other tanks, a graph can be constructed such that, the rate of ventilation of a tank compartment being known, the amount of HCN to be liberated to produce a lethal concentration can be directly estimated.

33. The results of attacks with Jet A.T. Mk.I charged HCN indicate that this German tank would need a ventilation rate of at least 10 changes of air per minute to render it reasonably safe from chemical weapon attack.

SIGNED

Air Commodore G. Combe, Chief Superintendent Porton Down.

Grenade, hand, lachrymatory (1lb) charged CAP bursting on Panzer III. (Z + 1/16 second).

Grenade, hand, lachrymatory (1lb) charged CAP bursting on Panzer III. (Z + 1/2 second).

Grenade, hand, lachrymatory (1lb) charged CAP bursting on Panzer III. (Z + 1 second).

.303/CAP bullets hitting Panzer III.

SECRET

VULNERABILITY OF GERMAN PZ.KW III TO ATTACK WITH SPECIALIZED ANTI-TANK CHEMICAL WEAPONS

REFERENCE

Priority Program Note M.211, Item 43(a, b, c and d).
Portion Report No.2444 Serial No.8, Date 29 Oct 1942

INTRODUCTION

1. This report describes the result of trials with specialized anti-tank chemical weapons against a German Pz.Kw III tank. The following weapons, which have hitherto been tried against British and American tanks only, were tested against the German armoured fighting vehicle:
   (i) Squirt, A.T. Portable Mk.I (charged HCN).
   (ii) Bomb, aircraft L.C. 65lb (charged vesicant substitute)
   (iii) Bomb, aircraft (experimental) (Charged HCN)
   (iv) Grenade, Hand, Lachrymatory (1lb) (Charged CAP)
   (v) .303/CAP. S.A.A.
2. A full examination of the ventilation of the tank was made and is the subject of a separate report.

DESCRIPTION OF THE PZ.KW III

3. The tank used in these trials was captured in the Libyan Campaign; it is of medium type, carries a crew of five, weighs 18 tons and has a top speed of about 25mph. A full examination of the actual tank used in the Porton trials was conducted by Messrs. A. E. C. Ltd, in June 1942 and a report can be obtained from the Director of Tank Design.
4. The features which would most affect a chemical warfare attack can be enumerated as follows:
      (i) Ventilation. The main ventilation of the interior of the tank is achieved in two ways:
   (a) All the air for the carburettors is drawn from the interior of the tank through air-cleaners, the inlets to which pass through a hatchway cover in the bulkhead between the engine and fighting compartments, and in turn the outside air is drawn into the interior through the gun mantlet and other interstices in the hull and turret.
   (b) The gun fumes are expelled by an extractor fan fitted in the roof.
      (ii) Hatches and Gun Mountings. The fighting compartment is very spacious and has a large number of hatches in both the welded hull

and the turret. These hatches are closely machined and are fitted with rubber gaskets where possible. The gun mountings are by no means liquid-proof but are protected by external armoured shields.

    (iii) Vision. Vision for the crew is through 'Schutzglaser' blocks and open slots; these being protected externally by armoured flaps operated from inside the tank.

    (iv) Armour. The armour of the tank is only 30mm on the front and sides, and as low as 12mm on the top (NB). It is understood that similar tanks now encountered in Libya have much thicker armour protection.

5. Most of the above features are shown in the drawing attached to this report.

6. When the tank reached Porton it was found that the hatchway cover in the engine bulkhead, the extractor fan in the roof and the front machine gun and mounting, were missing. A bulkhead, to the dimensions, was made and fitted; an extractor fan from a Churchill tank was satisfactorily fitted in the roof and the hole for the front gun was blanked off.

ATTACK WITH SQUIRT A.T. PORTABLE MK.I CHARGED HCN (PRUSSIC ACID)

Procedure

7. Nine separate attacks were made, and with exception of one attack when the Squirt was fired at the side of the tank, the Squirts were functioned at the front of the tank from a position about 20 yards upwind. The wind speed varied from 7 to 12 mph. On each occasion the tank was fully closed down, stationary, with engine running at about 1,700rpm and the extractor fan working.

8. As previously mentioned, the hatchway cover in the bulkhead between the engine and fighting compartment was missing and, as at first the principle of ventilation was not known, attacks with the Squirt were carried out (a) without the bulkhead, (b) with mock up bulkhead. When drawings had been obtained and a proper bulkhead made, further attacks with the Squirt were carried out.

Results

9. Detailed results of each of the nine attacks are recorded in the table attached to this report. The average concentrations (inside the tank) are above the dose which would produce fatal casualties amongst men unprotected by the respirator. In making this assessment the dosage of HCN required to cause immediate lethal effect in men has been taken as 10,000mg/cubic metre for a period of 30 seconds.

10. The concentrations recorded under the various conditions tend to show that the presence or absence of the hatchway cover in the engine bulkhead makes no significant difference to the effectiveness of the Squirt.

11. A photograph of the Squirt in action against the Pz.Kw III is attached.

ATTACK WITH BOMBS, AIRCRAFT L.C. 65LB

Procedure

12. The tank, closed down and without the engine running, was placed in a suitable position for low flying attack, facing into the wind. Bombs, aircraft L.C. 65lb charged with dyed water to represent the vesicant gas were dropped onto the front of the tank from a Lysander aircraft flying at 50 feet and at about 150 mph.

13. Observers dressed in white overalls and hoods were in the positions in the tank normally occupied by the crew.

Results

14. A good hit just below the mounting of the 50mm gun was obtained, and liquid penetrated, and splashed the interior of the tank in gross quantities.

15. The commander, front gunner, and loader were heavily splashed. The 50mm gunner and the driver escaped with minor contamination. It appeared that the main entry of the liquid was through the gun mantlet and the turret ring.

16. Previous trials with mustard gas on tanks have shown that if the liquid vesicant penetrated a tank in gross quantities, unprotected men would be eye and lung casualties from vapour alone as a result of a few minutes exposure, and those heavily splashed with liquid could only save themselves by evacuating the tank, removing all clothing, and applying anti-gas ointment to the skin without delay.

Panzer III after air attack with a 65lb chemical weapon bomb.

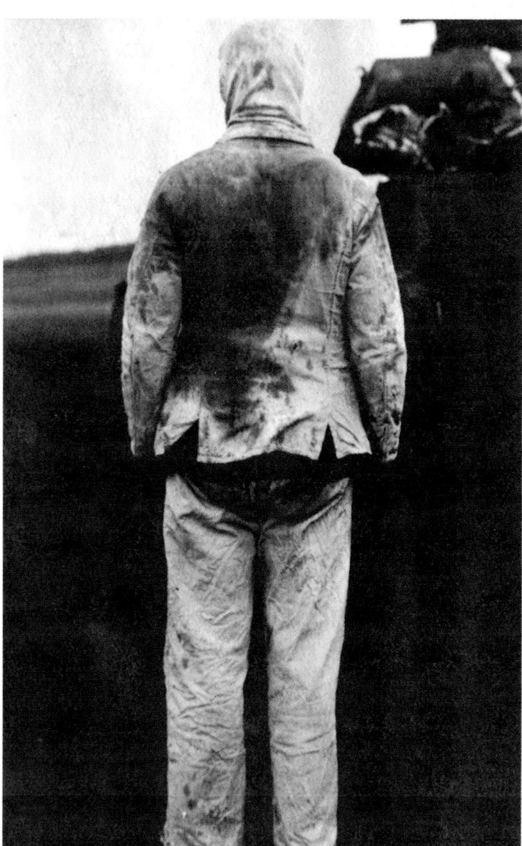

Above: The dark marks on the front of these men's clothing show where they would have been covered in chemicals during a bomb attack.

Left: Back of the Panzer III tank gun loader. The dark marks on his back and trousers show where he would have been covered in chemicals during a bomb attack.

ATTACK WITH EXPERIMENTAL BOMBS CHARGED WITH HCN

Procedure

Bombs similar to the 65 LB type, but cylindrical in Shape, and of heavier gauge metal to stand the vapour pressure of HCN, were dropped on the Pz.Kw III in the same manner as described in paragraph 12, with the exception that on this occasion the engine was set to run at 1,500rpm and the extractor fan in the roof was switched on. Guinea pigs were placed in all the positions.

Results

18. The aircraft obtained a good hit on the front of the tank. All the animals in the tank were found to be dead but owing to damage of the chemical sampling apparatus no reliable concentrations were obtained. The trial was repeated.

19. Three more bombs were dropped on the tank. The first missed, the second hit the ground in front of the tank, and enveloped the tank in HCN, while the third scored a glancing hit on the turret. The second and third bombs were sampled and concentrations of HCN inside the tank were as follows:

Time in seconds (Z = impact)	Concentrations (mg/m3)			
	2nd Bomb (Short miss)		3rd Bomb (Hit on Turret)	
	Driver's position	Commander's position	Driver's position	Commander's position
Z + 5	7,790	32,410	1,390	10,200
Z + 10	10,000*	18,440	520	Sample lost
Z + 20	20,000*	Sample lost	1,680	15,300
Z + 40	42,000	11,040	5,540	7,300
Z + 60	31,000	6,060	6,390	7,530
Z + 120	12,950	1,880	7,210	2,090

* These figures are approximate. Only in the former of these two attacks were animals used; they all died.

20. It is interesting to note from the above concentrations that a 'short miss' (bomb breaking up on ground in front of the tank) is more effective than the glancing hit on the turret. Both attacks would have been lethal to unprotected men in the tank.

ATTACK WITH GRENADE, HAND, LACHRYMATORY (1LB) CHARGED C.A.P.

Procedure

21. In order to reproduce the standard conditions under which this grenade has been tested against British tanks, one grenade was strapped onto the front of

the tank just below the gun mantlet and functioned electrically. The tank with engine running at 1,500rpm and extractor fan on, was stationary and facing into the wind. The wind, speed was 6mph. Chemical samples were taken in the driver's and commander's positions, but no observers were in the tank on this occasion.

Result

22. The grenade functioned correctly and formed a heavy cloud of CAP round the turret of the tank.

23. The following concentrations of CAP were recorded:

Time in seconds (Z = impact)	Concentrations in mg.CAP/m^3		Physiological Interpretation
	Driver's position	Commander's position	
Z + 2	Sample lost	22	Concentrations of this order have proved sufficient in British tanks to cause instantaneous incapacitation of crew.
Z + 5	20	49	
Z + 10	17	36	
Z + 20	16	24	
Z + 30	20	21	
Z + 60	9	12	

ATTACK WITH .303/CAP BULLETS

Procedure

24. Twenty-eight rounds (one magazine) of .303/CAP G Mk.II ammunition were fired in a single burst at 250 yards range at the front of the Pz.Kw III. The tank was stationary with the engine running and extractor fan on. No observers were in the tank as the danger from lead splash was not known, but chemical samples were taken.

Results

25. All the rounds hit the tank in the region of the driver's visor and the base of the gun mounting. A small cloud of CAP enveloped the turret of the tank and concentrations of between 5 and 21mg/m^3 were recorded in the fighting compartment.

26. Previous trials with this ammunition, fired at British tanks, have shown that temporary incapacitation due to Lachrymation, will interfere with the normal functions of the crew and possibly cause the tank to be halted. The concentrations recorded show that the German Pz.Kw III is equally as vulnerable to this form of attack as the British tanks on which trials have been carried out.

CONCLUSIONS

27. The special chemical weapons that have been designed for attack on tanks are effective against the German Pz.Kw III and the concentrations recorded are of the same order as those obtained on the British and American tanks which had been used as a standard during the development of the weapons.

SIGNED

Air Commodore G. Combe, Chief Superintendent

SQUIRT ANTI-TANK PORTABLE MK.I CHARGED HCN

Concentrations in milligrams HCN per cubic metre—Z = time of HCN jet hitting the tank

Attack No.	Special conditions.	Time in seconds	Concentrations		Effect on animals in tank	Remarks
			Driver's position	Commander's position		
1	Mock-up bulkhead in position.	Z + 5	390	Sample Lost	All rabbits in crew positions died but goat in turret remained alive but stupefied.	This attack can be regarded as a near-miss as most of the HCN did not hit the tank.
		Z + 10	650	1,310		
		Z + 20	2,890	780		
		Z + 45	Sample Lost	40?		
		Z + 60	2,310	460		
		Z + 120	590	40		
2	Bulkhead removed.	Z + 5	28,500	Sample Lost	All the animals, 1 goat and 3 rabbits, died.	A good hit—tank wet with HCN. An unprotected crew would almost certainly have been fatal casualties.
		Z + 10	18,650	12,750		
		Z + 20	21,200	14,000		
		Z + 45	Sample Lost	2,790		
		Z + 60	10,900	Sample Lost		
		Z + 120	5,160	790		
3	Bulkhead removed.	Z + 5	Sample Lost	14,100	No animals used but had they been all would have died.	Ditto.
		Z + 10	Sample Lost	10,300		
		Z + 20	14,200	13,350		
		Z + 45	Sample Lost	6,900		
		Z + 60	17,200	3,800		
		Z + 120	4,200	1,000		
4	Bulkhead removed.	Z + 5	3,950	20,000	No animals used but had they been all would have died.	Ditto.
		Z + 10	14,800	13,400		
		Z + 20	24,700	17,700		
		Z + 45	16,100	Sample Lost		
		Z + 60	13,600	390		
		Z + 120	4,150	390		

Attack No.	Special conditions.	Time in seconds	Concentrations		Effect on animals in tank	Remarks
			Driver's position	Commander's position		
5	Mock-up bulkhead in position.	Z + 5 Z + 10 Z + 20 Z + 45 Z + 60 Z + 120	4,500 7,000 27,000 11,900 10,050 Sample Lost	6,450 8,250 6,500 1,100 600 160	No animals used but had they been all would have died.	Ditto.
6	Mock-up bulkhead in position.	Z + 5 Z + 10 Z + 20 Z + 45 Z + 60 Z + 120	Sample Lost Sample Lost 15,700 8,300 13,500 4,900	16,600 Sample Lost 10,100 Sample Lost 2,800 790	No animals used but had they been all would have died.	Ditto.
7	Correct ventilation system fitted.	Z + 5 Z + 10 Z + 20 Z + 45 Z + 60 Z + 120	740 Sample Lost 8,400 4,610 2,960 195	5,00 8,480 Sample Lost 3,230 1,910 780	A goat, rabbit and guinea pigs placed in the crew positions all died	These concentrations are rather lower than previously recorded, but would certainly produce unconsciousness in unprotected human beings.
8	Correct ventilation system fitted.	Z + 5 Z + 10 Z + 20 Z + 45 Z + 60 Z + 120	3,520 Sample Lost 11,800 10,530 3,440 3,330	7,660 9,060 Sample Lost 8,430 5,710 1,620	A goat, rabbit and guinea pigs placed in the crew positions all died	Having the correct ventilation system appeared to make very little difference to the concentrations of HCN.
9	Correct ventilation system fitted. Tank attacked from the side.	Z + 5 Z + 10 Z + 20 Z + 45 Z + 60 Z + 120	1,350 12,740 22,500 20,610 12,830 2,930	11,580 7,500 6,000 4,600 1,170 Sample Lost	Six guinea pigs in crew positions all died.	This proves that a side hit on the tank is as effective as a hit on the front.

Attack No.	Special conditions.	Time in seconds	Concentrations		Effect on animals in tank	Remarks
			Driver's position	Commander's position		
Average concentrations under all conditions.		Z + 5	6,130,	11,690		
		Z + 10	10,770	9,070		
		Z + 20	16,480	9,770		
		Z + 45	12,010	4,510		
		Z + 60	9,640	2,100		
		Z + 120	3,180	690		
	Over 1min		11,060	7,430		

REMARKS

(i) It will be noticed that the concentrations reach their peak in the driver's position at Z + 20 secs whereas in the commander's position they fall rapidly after the initial concentration.

(ii) The mean concentrations recorded would be lethal to unprotected personnel if breathed for a few seconds.

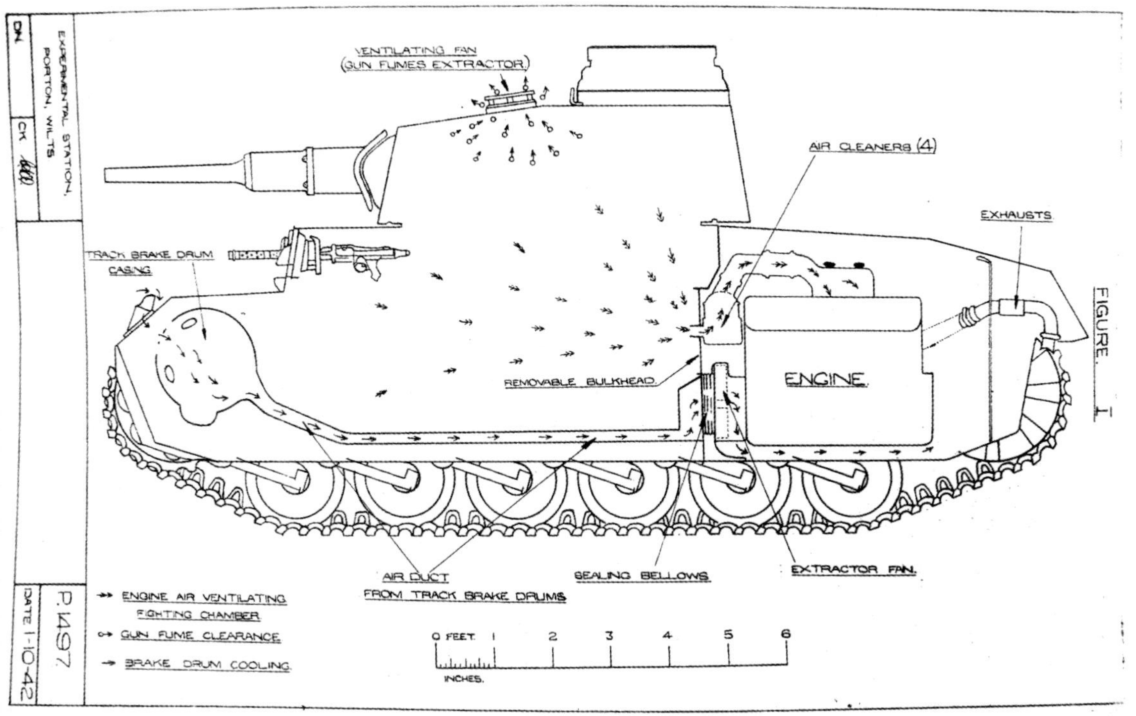

Air flow diagram inside a Panzer III tank.

4

Panzerkampfwagen IV Ausführung D

The long version of this tanks name is Panzerkampfwagen-IV (7,5cm) (Vs.Kfz. 622) Ausführung D, (4. und 5.Serie/B.W.) (Sd. Kfz. 161). Of the 248 ordered Panzer IV Ausf.D tanks, a total of 231 were completed between October 1938 and October 1939. The order for the first 200 was called series 4 (4. /B.W.),

Panzerkampfwagen IV Ausführung D with turret number 813 was not up armoured when it arrived in North Africa. 5th Panzer Regiment, 21st Panzer Division, Afrikakorps, 1941. Its front armour was only 30mm thick.

and the further 48 were called series 5 (5. /B.W.). Both series 4 and 5 /BW were of the same design.

Some of the remaining nineteen hulls were used for special versions: sixteen were used to construct Bruckenleger IV tanks (armoured-vehicle bridge laying tanks); two for the 10.5cm K18 Sf.IV Dicker Max self-propelled gun and one as an ammunition carrier for the Karl-Gerät, a super-heavy mortar. One tank was used in the trials to up-gun the Pz.Kpfw. IV with high-velocity guns. It was equipped with a 5cm Kw.K.39 L/60.

A front hull armoured ball mounted 7.92mm M.G.34 machine gun was reintroduced. The driver's front was stepped forwards, similar to the Panzer IV Ausf.A, with a circular vision/pistol port added in the resulting central corner. This gave the driver more vision to his right.

The 7.5cm Kw.K L/24 main gun mantlet was reinforced with a slightly curved armour plate of 35mm thickness. The side and rear armour of the Ausf.D were increased from 14.5mm to 20mm, somewhat improving its survivability.

The front hull and superstructure were built with 30mm-thick face-hardened armour. In February 1940, 30mm-thick applique armour plates were bolted or welded to the front superstructure and hull, bringing the armour protection up to 60mm thick in these areas. Also, 20mm applique armour plates were bolted or welded to the sides, increasing the side armour in the centre to 40mm. The last 68 Panzer IV Ausf.D tanks had 50mm thick front hull armour instead of the original 30mm. The increased thickness of the armour increased the weight of the Panzer IV Ausf.D to 20 tonnes. What follows is the British report on the examination of a captured German Panzer IV Ausf.D tank.

DEPARTMENT OF TANK DESIGN WO 194/181

D.T.D EXPERIMENTAL REPORT F.T. 214

PANZERKAMPFWAGEN IV AUSFÜHRUNG D EXAMINATION

ORIGIN

This tank was captured from the enemy in the Middle East, [believed 14 April 1941] and it was sent home for information to be obtained regarding construction and performance.

REMARKS

The machine was received in so damaged a state that no running trials have been possible. This report is the result of an external examination only.

FURTHER ACTION

The tank has now been stripped for examination and repair by Leyland motors. It is intended that performance trials shall be carried out when the tank returns to Farnborough.

STATEMENT

The German tank PZKW 4 was delivered to the experimental wing on third of November 1941. Whilst the machine was on board ship in Port Said it was damaged by enemy action. It was, however, a 'runner' when it left Tobruk, where it was captured. Fire broke out on the ship and the damage was caused to the vehicle by fire, water and fire foam. All the information contained in this report was obtained without dismantling any parts.

OBSERVATIONS

Condition

All the linkages in the break actuation system, steering control rods, turnbuckles, door hinges, vizor flaps and locks, et cetera are so badly rusted that no movement can be made without damaging them. The number three offside bogey assembly and bracket is missing. The rubber tyres on the two rear offside guide rollers and number four offside bogey assembly have been burnt off. The armour plate in front of the gearbox was holed by some type of high explosive round whilst the machine was in action, and the horizontal plate which mates with the plate holding the hull gunner's look-out has been dented by some explosive, probably a bomb.

The clutch housing (just before the gearbox) has been damaged by some type of hand grenade, no doubt put near there by the crew in their efforts to destroy the wireless set and other useful instruments mounted immediately above. Judging from a cursory inspection from the outside without dismantling any of the parts, no damage appears to have been done to the clutch plates. The majority of the electrical wiring system has been destroyed by being pulled away from its sockets and mountings with the exception of the armoured cable which has been left in situ.

The turret is jammed by rust and dirt but can be moved slightly by the use of the hand traversing handle with external assistance. The auxiliary charging and power traverse unit is badly rusted externally. The power unit, a V-12, 320HP Nordbau engine, appears to be in a fair condition with the exception of moderately heavy deposits of rust round the block and sump. All the stowage; field glasses, telescopes, ammunition, Verey light pistols, machine guns and suchlike equipment, had been removed before the vehicle arrived.

Vehicle characteristics

The weight of the vehicle quoted on the bill of landing was 22 tons and agrees with the weight quoted in past intelligent summaries. One of the first impressions an observer of this tank gets is that the whole machine, from the hull to the smallest detail, is fabricated. The hull and turret are all welded construction of flat rolled plates with a few minor exceptions such as the gun mantlet, the look-out masks and the sprocket hubs.

In plain, the lower half of the hull is rectangular in shape, 6ft 3in wide and 16ft 10in in length, the side plates running the full length of the vehicle. This half of the hull—up to as far as the mudguards, is made independently of the top half. The two are joined together by a strip of angle iron on either side which is tack welded along the bottom edge, the top half being bolted and split pinned to the other flange.

The vehicle is carried on eight cast mild steel brackets—four on either side, in each of which are mounted two cast cranked levers, with the outer ends facing fore and aft. The forward lever carries a twin bogey wheel 18½ inch in diameter mounted on two heavy duty ball races and a cage integral with the arm, in which is anchored the rigid end of a true cantilever quarter elliptic spring. This in turn is held within the cage by two ¾in bolts passing through the cage and the leaves, the nuts and split pins being on the underside. The other arm is fitted with the second twin bogey wheel and a roller shackle for the free end of the spring.

Any upward or downward movement of either wheel in an assembly slides the trailing end of the spring in the roller shackle whilst the spring itself is being deflected. The movement on the fixed end of the spring is on the ark of the bogey arm only.

The bumper pad system is interesting, in that the upward movement of the leading wheel is controlled by a pad mounted immediately above it, whilst the upward movement of the second wheel, i.e. each wheel of each bogey assembly which carries the trailing end of the spring in a roller shackle, is controlled by the first wheel of the bogey assembly after it. The efficiency of this arrangement is brought about by the small clearances between each wheel and the small angle the track will make between each wheel whilst climbing over small vertical objects. An exception to the above is to be found on the last bogie assembly which has a pad for each arm, as the idler which follows it is off the ground and further away.

The vizor flaps are spigotted to fit snugly into the aperture leaving a large flange all around. No refinements such as periscopes are fitted, all observation being carried out from within through either slits or 'Ersatzgläser' glass vision blocks, these being small arms proof.

All the external heads of rivets and the locking devices of the revolver ports, fuel tank and fan belt adjustment covers, are highly conical. The difference between the rivets and locks being two small flats cut on either side for a suitable sized spanner.

The smoke candle rack for five candles is mounted on the exhaust system. The smoke candle is ejected from the rack by a spring and is automatically ignited. Five compartments are controlled by a camshaft with the cams placed round the shaft at 72°. On the end of the shaft is a ratchet toothed wheel with the same angular spacing; this is operated by a spring-loaded pull rod by the tank commander, one candle being released for each pull.

The general exterior measurements are	
Overall width (edges of mudguards)	9ft 2 ¼ in
Overall length (excluding mudguards, as the ends are hinged)	18ft 8 ½ in
Hull length	16ft 8 ½ in
Hull width, lower half	6ft 3in
Hull width, upper half	7ft 8 ¼ in
Overall height	8ft 3 ½ in
Ground clearance	1ft 4 ¼ in

Track data	
Length of track on the ground	11ft 6 ¾ in
Overall length of track	18ft 5in
Track centres	7ft 10 ¾ in
Track width	1ft 2in
Track plate height to top of guide tongue	5 ¼ in
Guide tongue height only	3 $^1/_8$in
Pitch of track	4.76in
Track pin length to head	1ft 2 ¾ in
Track pin length overall	1ft 3in
Track pin diameter	¾ in
Weight of track plate and pin	14 lb 9.1oz
Number of plates per track	98
Diameter of idler	2ft 3 ½ in
Centre of idler from ground	1ft 10 $^3/_8$in
Diameter of sprockets (tops of teeth)	2ft 7 ¼ in
Diameter of sprockets (bottom of teeth)	2ft 3 ½ in
Pitch diameter	2ft 5 ½ in
Height of sprockets centre from ground	2ft 4 ½ in
Number of teeth on sprocket	19
Number of twin bogey wheels per side	8
Diameter of twin bogey wheels per side	1ft 6 ½ in
Depth of rubber	1 ¾ in
Width (one side of twin wheel only) at base	2 ¾ in
Width (one side of twin wheel only) to top	2 $^3/_8$in

Quarter elliptic springs	
Number of leaves	14
Thickness of leaves	$^3/_8$in
Width of leaves	3 ½ in
Length of main leaf (approximate)	2ft 8in

Each bogey wheel is mounted on ball bearings and filled with a type of mineral jelly. The sprocket is designed so that the sprocket rings can be changed without having to dismantle it. The inside of the sprocket ring has tongues, which have to mate with a corresponding external type on the sprocket hub. The space between these external tongues is recessed so that the inner sprocket ring tongues can be passed over the mounting then rotated through an angle of about 10° to line up the bolt holes. Two small circular hinged plates are fitted into the nearside armour plate, under the track, to cover the entrances to the petrol tanks. Armour plate thicknesses are:

All vertical front plates	35mm
Sloping front plates	20mm
Horizontal plates	10mm
Side plates, hull and turret	20mm
Back plates	20mm

The turret base is protected by an inverted 'V' section ring bolted to the hull top plating. This, however, does not completely encompass the turret, the back and nearside being unprotected. The reason for this Curia formation is that the turret is offset to the nearside by approximately 2 inches.

The whole of each side above the mudguards by the engine compartment is louvred and fitted with small simple flat shutters for varying the cooling air inlet and outlet flow. In addition to these movers, which can be shut off altogether, are two small auxiliary ones on the top plating made by rectangular slots cut in the plating with slightly larger similar shaped plates fitted ½ inch above and all enclosed round the edge by an inch wall of 10mm plating. These louvres are permanently open and are vulnerable to either high explosive bomb shrapnel or ball ammunition fired from aircraft.

The majority of the space on top of the mudguards is used for carrying spare track plates, about 30 in number. Two spare bogie wheels are carried on brackets mounted on either side of the rear of the vehicle.

Interior

The engine compartment is in the rear with the main power unit, a 320 H.P. Nordbau motor, mounted on the offside. On the nearside is a small two-stroke

2 H.P. DKW auxiliary motor with flywheel ignition for a charging unit and power traverse dynamo. Both these engines are air cooled.

The main power unit is fitted with twin carburettors mounted behind the two banks with a single air inlet pipe. The air is drawn from a hole in the offside armour plate near the rear of the turret. An air duct takes the air down into the air cleaner, mounted on the offside of the engine via a manually operated valve, so that the air can be drawn from either side, from the fighting compartment or both.

The ignition is by a Bosch magneto mounted under the inlet pipe plugs mounted on the inner side of each bank and are served by rubber fabric insulated leads. These are covered in by an aluminium cover which serves as a screen to prevent wireless interference. Dual ignition is not provided. The overhead gear driven camshaft runs down the centre of each head with the rocker shaft mounted on either side of it. The inlet valves being operated by the rocker arms mounted on the upper shaft, whilst the exit valves are operated by the rocker arms mounted on the lower shaft. These being transmitted from the cam through rollers mounted in the rocker arm. There are two methods of starting: two electrical starter motors or by using the starting handle, this having to be used on the outside of the tank.

Transmission

The propeller shaft is mounted direct onto the flywheel and passes under the turntable of the turret to a transmission fan which draws air over the steering brake mechanism, blowing it through a pipe mounted on a central column and passing behind the driver's head; the air is finally expelled through a semi-circular cowl mounted on the nearside of the hull. The shaft then passes into a multi-plate clutch mounted on the gearbox. This arrangement shortening the lengthy control rods similar to those used on our machines.

The gearbox is thought to be an 8-speed Maybach box. The drive emerges from the box via a bevel, the cross shaft passing close to the front plating. There are two Layrub type couplings mounted on each shaft before the sprocket.

Steering is thought to be affected by a 2-stage epicyclic, mounted on each side, with brake shoes fitted with cast iron linings to affect in the necessary breaking of the annuli. Oil is pressure fed from the gearbox to these steering units and is returned by a second set of pipes. To the right of the driver, who sits on the nearside of the gearbox, is the remains of an instrument panel, but there is no indication of the type of instruments fitted. There is no trace of any of the following: speedometer, revolution counter, clock, oil gauge, fuel gauge or ammeter.

It is assumed that the intercommunication was electrical and worked in conjunction with the wireless set, but as the whole of the electrical system forward is damaged, and there is nothing to be found in the nature of headphones, loudspeakers or hand microphones, it is hard to give an indication of the type used.

Armament

The tank should be fitted with two machine guns of the Rheinmetall type, one forward operated by the wireless operator on the right of the driver, and a co-axle gun, pedal operated by the gunner. The main armament is a 75mm low muzzle velocity gun fired electrically by a finger trigger mounted on the hand traversing handle.

Intelligence summaries have stated that the vehicle carries 84 rounds for this gun, but stowage bins for only 44 can be found. Electrical power traverse is fitted to the left of the 75mm gun and is provided with the necessary current by the small 2 H.P. 2-stroke D.Kw (Auto Union) petrol engine driving a dynamo. 25 gallons of petrol are carried for the use of this unit. The operation of the power traverse unit seems to be by a control lever mounted on the top of the gearbox, which has three positions; right, neutral, left. This method gives one speed only to the right or left and leaves the gunner to do the fine laying by the hand traversing handle.

The size of this unit is enormous for the job it has to perform and is the only fitting which is incongruous with the rest of the vehicle. Although no telescope was fitted when it was delivered, the size of the power traverse gear indicates that it must be between 2ft 6 inch and 3ft long.

Seats are provided for every member of the crew and it is unlikely that anyone, other than the tank commander, who is provided with a cupola, would stand for any length of time. The head room provided in the turret is approximately 4ft 3 inch. This is no doubt due to two reasons: the propeller shaft passing under the turntable and having to keep the height of the vehicle within reasonable limits.

The turret ring is 64 inches in diameter and is larger than any fitted to our vehicles. This is responsible for the 'roominess' of the fighting compartment. The tank commander's cupola is fitted with five vizors giving him an all-round field of view, but with rather a restricted view of the ground close to the tank, the nearest point being approximate 14 yards away. With the tank completely closed down it is blind on three sides at very close range in spite of the fact that it is provided with 13 vizors or look-outs. The gun and the hull gunner are the only members of the crew who can see the ground a short distance in front, their vision to the right and left being as poor as that of the others.

Six escape hatches are provided; on top of the cupola, the right side of the turret, the left side of the turret, above the driver's head, above the wireless operator's head, and below the wireless operator's seat, the seat being hinged so they can be folded forward.

Fire control

To the left of the gunner is mounted a small dial numbered from 1 to 12. It is driven by a small shaft geared to the turret rack. Another gear at the same size drives a ring in the cupola, which is also numbered from 1 to 12. On the outside of the cupola is a small pointer which is just visible through the look-out. The figure opposite the pointer on the cupola scale corresponds with a figure opposite the small pointer on the gunner's scale. If the commander were to see a target through one of the other lookouts, he would call the number immediately above that look-out in which the target appeared. The gunner would then traverse until that figure came up to his pointer. At that stage he has been practically laid on to his target, except for lateral corrections and elevation.

SIGNED

Officer in command of Field Trials, Department of Tank Design, 2 December 1941.

DEPARTMENT OF TANK DESIGN

FIELD TRIALS REPORT ON GERMAN TANK PANZER IV

STATEMENT

The German tank Panzer IV was delivered to the Experimental Wing on 3 November 1941. Whilst the machine was on board ship in Port Said it was damaged by enemy action. It was however, a 'runner' when it left Tobruk, where it was captured. Fire broke out on the ship and damage was caused to the vehicle by: (a) Fire, (b) Water, (c) Fire foam. All information contained in this report was obtained without dismantling any parts.

OBSERVATIONS

Conditions

1. All the linkages in the brake actuation system, steering control rods, turn buckles, door hinges, vizor flaps and locks, etc, are so badly rusted that no movement can be made without damaging them.
2.
    (a) No.3 O.S. bogie assembly and bracket are missing.
    (b) The rubber tyres on the two rear O.S. guide rollers and No.4 O.S. bogie assembly have been burnt off.
3. The armour plate in front of the gearbox was holed by some type of HE whilst the machine was in action, and the horizontal plate which mates with the plate holding the hull gunner's look-out has been dented by some explosives, presumably a bomb.
4. The clutch housing (just before the gearbox) has been damaged by some type of hand grenade, no doubt put near there by the crew in their efforts to destroy the wireless set and other useful instruments mounted immediately above. Judging from a cursory inspection from the outside without dismantling any of the parts, no damage appears to have been done to the clutch plates.

5. The majority of the electrical wiring system has been destroyed by being pulled away from its sockets and mountings with the exception of the armoured cable which has been left in situ.
6. The turret is jammed by rust and dirt but can be moved slightly by the use of the hand traversing handle with external assistance.
7. The auxiliary charging and power traverse unit in badly rusted externally.
8. The power unit, a V-12, 320hp Nordbau engine, appears to be in a fair condition with the exception of moderately heavy deposits of rust round the block and sump.
9. All the stowage; field glasses, telescopes, ammunition, Verey light pistols, machine guns and such like equipment, had been removed before the vehicle arrived at E.W.

Vehicle characteristics

(a) The weight of the vehicle quoted on the Bill of Landing was 22 tons and agrees with the weight quoted in past Intelligence Summaries.

(b) One of the first impressions an observer of this tank gets is that the whole of the machine, from the hull to the smallest detail, is fabricated. The hull and the turret are of all-welded construction of flat rolled plates with a few minor exceptions such as the gun mantlet, the look-out masks and the sprocket hubs. In plan the lower half of the hull is rectangular in shape, 6ft 3in wide and 16ft 10in in length, the side plates running the full length of the vehicle. This half of the hull—up to as far as the mudguards, is made independently of the top half. The two are joined together by a strip of angle iron on either side which is tack welded along the bottom edge, the top half being bolted and split pinned to the other flange.

(c) The vehicle is carried on eight cast mild steel brackets—four on either side, in each of which are mounted two cast cranked levers, with the outer ends facing fore and aft. The forward lever carries (1) a twin bogie wheel 18 ½ in diameter mounted on two heavy duty ball races and (2) a cage integral with the arm in which is anchored the rigid end of a true cantilever quarter elliptic spring. This in turn is held within the cage by two ¾ in bolts passing through the cage and the leaves, the nuts and split pins being on the underside.

The other arm is fitted with the 2nd twin bogie wheel and a roller shackle for the free end of the spring. Any upward or downward movement of either wheel in an assembly slides the trailing end of the spring in the roller shackle whilst the spring itself is being deflected. The movement on the fixed end of the spring is on the arc of the bogie arm only.

The bumper pad system is interesting, in that the upward movement of the leading wheel is controlled by a pad mounted immediately above it, whilst the upward movement of the second wheel, i.e. each wheel of each bogie assembly which carries the trailing end of the spring in a roller shackle, is controlled by the first wheel of the bogie assembly after it. The efficacy of this arrangement is brought about by the small clearance between each wheel and the small angle the track will make between each wheel whilst climbing over small vertical objects.

An exception to the above is to be found on the last bogie assembly which has a pad for each arm, as the idler which follows it is off the ground and further away.

(d) The vizor flaps are spigotted to fit snugly into the aperture leaving a large flange all round. No refinements such as periscopes are fitted, all observations being carried out from within through either slits or 'Erasatzgläser' blocks, these being small arms proof.

(e) All the external heads of rivets and locking devices of revolver ports, fuel tank and fan belt adjustment covers, are highly conical. The difference between the rivets and locks being two small flats cut on either side for a suitably sized spanner.

(f) A smoke candle rack for 5 candles is mounted on the exhaust system. The smoke candle is ejected from the rack by a spring and is automatically ignited. The five compartments are controlled by a camshaft with the cams placed round the shaft at 72°. On the end of this shaft is a ratchet toothed wheel with the same angular spacing; this is operated by a spring-loaded pull rod by the tank commander, one candle being released for each pull.

(g) The general exterior measurements are:

Overall width (edge of mudguards)	9ft 2 ¼in
Overall length (excluding mudguards, as the ends are hinged)	18ft 8 ½in
Hull length	16ft 10in
Hull width, lower half	6ft 3in
Hull width, upper half	7ft 8 ¼in
Overall height	8ft 3 ½in
Ground clearance	1ft 4 ¼in

Track data	
Length of track on the ground	11ft 6 ¾in
Overall length of track	18ft 5in
Track centres	7ft 10 ¾in
Track width	1ft 2in
Track plate height to top of guide tongue	5 ¼in
Guide tongue height only	3 ⅛in
Pitch of track	4.76in
Track pin length to head	1ft 2 ¾in

Track data	
Track pin length to overall	Ift 3in
Track pin diameter	¾ in
Weight of track plate and pin	I4lb 9.1oz
Diameter of idler	2ft 3 ½in
Centre of idler from ground	Ift I0 ³/₈in
Diameter of sprocket (tops of teeth)	2ft 7 ¼in
Diameter of sprocket (bottom of teeth)	2ft 3 ½in
Pitch diameter	2ft 5 ½in
Height of sprocket centre from ground	2ft 4 ½in
Number of teeth on sprocket	I9
Number of twin bogie wheels per side	8
Diameter of twin bogie wheels per side	Ift 6 ½in
Depth of rubber	I ¾in
Width (one side of twin wheel only) at base	2 ¾in
Width (one side of twin wheel only) at top	2 ³/₈in
Quarter elliptic springs, number of leaves	I4
Quarter elliptic springs, thickness of leaves	³/₈in
Quarter elliptic springs, width of leaves	3 ½in
Quarter elliptic springs, length of main leaf (approx.:)	2ft 8in

(h) Each bogie wheel is mounted on ball bearings and filled with a type of mineral jelly.

(i) The sprocket is designed so that the sprocket rings can be changed without having to dismantle it. The inside of the sprocket ring has tongues, which have to mate a corresponding external type on the sprocket hub. The space between these external tongues is recessed so that the inner sprocket ring tongues can be passed over the mounting then rotated through an angle of about I0° to line up the bolt holes.

(j) Two small circular hinged plates are fitted into the N.S. armour plate under the track to cover the entrance to the petrol tanks.

(k) The armour plate thicknesses are:

All vertical front plates	35mm
Sloping front plates	20mm
Horizontal plates	I0mm
Side plates, hull and turret	20mm
Back plates	20mm

(l) The turret base is protected by an inverted 'V' section ring bolted to the hull top plating. This, however, does not completely encompass the turret, the back and N.S. being unprotected. The reason for this peculiar formation is that the turret is off set to the N.S. by approximately 2 inches.

(m) The whole of each side above the mudguards by the engine compartment is louvered and fitted with small simple flat shutters for varying the cooling air inlet and outlet flow. In addition to these louvres, which can be shut off though, are two small auxiliary ones on the top plating made by rectangular slots cut in the plating with slightly larger similarly shaped plates fitted ½ inch above and all enclosed round the edge by an inch wall of 10mm plating. These louvres are permanently open and are vulnerable to either A.W. bombs or ball ammunition fired from aircraft.

(n) The majority of the space on the top of the mudguards is used for carrying spare track plates, about 30 in number. Two spare bogie wheels are carried on brackets mounted on either side of the rear of the vehicle.

Interior

The engine compartment is in the rear with the main power unit 320 hp Nordbau motor, mounted on the off-side. On the near-side is a charging unit and power traverse dynamo. Both these engines are water cooled. The main power unit is fitted with twin carburettors mounted between the two banks with a single air inlet pipe. The air is drawn from a square hole in the O.S. armour plate near the rear of the turret. An air duct takes the air down into the air cleaner, mounted on the O.S. of the engine vis a manually operated valve, so that the air can be drawn either from a side, from the fighting compartment, or both.

Original report diagram showing air intake on the Panzer IV.

The ignition is by a Bosch magneto mounted under the inlet pipe. Spark plugs are mounted on the inner side of each bank and are served by rubber fabric insulated leads. These are covered in by an aluminium cover plan which serves as a screen to prevent wireless interference. Dual ignition is not provided.

The overhead-gear driven camshaft runs down the centre of each head with a rocker shaft mounted on either side of it. The inlet valves being operated by the rocker arms mounted on the upper shaft, whilst the exhaust valves are operated by the rocker arms mounted on the lower shaft. The most being transmitted from the cam through rollers mounted in the rocker arms.

There are two methods for starting:

1. Two electric starter motors.
2. Starting handle; this having to be used on the outside of the tank.

Transmission

The propellor shaft is mounted direct on to the flywheel and passes under the turn-table of the turret to a transmission fan which draws air over the steering braking mechanism, blowing it through a pipe mounted on a central column and passing behind the driver's head; the air is finally expelled through a semi-circular cowl mounted on the N.S. of the hull just forward of the turret.

The shaft then passes into a multi-plate clutch mounted on the gearbox, this arrangement shortening the lengthy control rods similar to those used in our machines. The gearbox is thought to be an 8-speed Maybach box. The drive emerges from the box via a bevel, the cross shaft passing close to the front plating. There are two Layrub type couplings mounted on each shaft before the sprocket.

Steering is thought to be affected by a 2-stage epicyclic, mounted on each side, with brake shoes fitted with cast iron linings to affect the necessary braking of the annuli. Oil is pressure fed from the gearbox to these steering units and is returned by a second set of pipes.

To the right of the driver, who sits on the near side of the gearbox, is the remains of an instrument panel, but there is no indication of the types of instruments fitted. There is no trace of any of the following necessaries:

1. Speedometer
2. Revolution counter
3. Clock
4. Oil gauge
5. Fuel gauge
6. Electric meter

It is assumed that the intercommunication was electrical and worked in conjunction with the wireless set, but as the whole of the electrical system forward is damaged, and there is nothing to be found in the nature of headphones loudspeakers or hand microphones, it is hard to give an indication of the type used.

Armament

The tank should be fitted with two machine guns of the Rheinmetall type, (a) one forward, operated by the wireless operator on the right of the driver, and (b) one co-axial gun pedal operated by the gunner. The main armament is a 75mm low muzzle velocity gun fired electrically by a finger trigger mounted on the hand traversing handle.

Intelligence Summaries have stated that the vehicle carries 74 rounds for this gun, but stowage bins for only 44 can be found. Electrical power traverse is fitted to the left of the 75mm gun and is provided with the necessary current by the small 2hp 2-stroke D.Kw (Auto Union) petrol engine driving a dynamo. Twenty-five gallons of petrol are carried for the use of this unit.

The operation of the power traverse unit seems to be by a control lever mounted on the top of the gearbox, which has three positions: - (a) right, (b) neutral, (c) left. This method gives one speed only to the right or left and leaves the gunner to do the fine laying by the hand traversing handle. The size of this unit is enormous for the job it has to perform and is the only fitting which is incongruous with the rest of the vehicle. Although no telescope was fitted when it was delivered to E.W. the size of the power traverse gear indicates that it must be between 2ft 6in and 3ft long.

Seats are provided for every member of the crew and it is unlikely that anyone, other than the tank commander, who is provided with a cupola, would stand for any length of time. The head room provided in the turret is approximately 4ft 3in. This is no doubt due to two reasons: (1) the propeller shaft passing under the turntable, and (2) having to keep the height of the vehicle within reasonable limits.

The turret ring is 64″ in diameter and is larger than any type fitted to our vehicles. This is responsible for the 'roominess' of the fighting compartment. The Tank Commander's cupola is fitted with 5 vizors giving him an all-round field of view, but with rather a restricted view of the ground close to the tank, the nearest point being approximately 14 yards away. With the tank completely closed down it is blind on three sides at very close range in spite of the fact that it is provided with 13 vizors or look-outs. The driver and the hull gunner are the only members of the crew who can see the ground a short distance on front, their vision to the right and left being as poor as that of the others.

Six escape hatches are provided:

1. The top of the cupola
2. The right side of the turret
3. The left side of the turret
4. Above the driver's head
5. Above the wireless operator's head
6. Below the wireless operator's seat, the seat being hinged so that it can be folded forward.

Fire control

To the left of the gunner is mounted a small dial number from 1 to 12. This is driven by a small shaft geared to the turret rack. Another gear of the same size drives a ring on the cupola, which is also numbered from 1 to 12. On the outside of the cupola is a small pointer which is just visible through the look-out. The figure opposite the pointer on the cupola scale corresponds with the figure opposite the small pointer on the gunner's scale. N.W. if the commander were to see a target through one of the other look-outs he would order the number immediately above that look-out in which the target appeared, the gunner would then traverse until that figure came up to the pointer. At that stage he has been practically laid on to his target, except for slight lateral corrections and elevation.

SIGNED

Department of Tank Design
Officer in charge of Field Trials
2 December 1941

OPERATIONAL HISTORY OF GERMAN PANZER IV TURRET NUMBER 813

This tank belonged to the German Africa Korps (DAK), 87th Company, Panzer Regiment 5, 5th Light Division later 21st Panzer Division. It was previously issued to the 3rd Panzer Division. In the hurry to get the tank shipped to North Africa it still displayed the 3rd Panzer Division identification sign, an inverted Y with two small vertical lines. The Panzer Regiment 5 was sent to form the 5th Light Division as an initial cadre of the DAK, the division later evolved into the 15th Panzer Division.

The regiment was provided with two Panzer IV companies, the 4th and the 8th, with 10 tanks each. Their numbering provides the first tank ID resource.

The 8th company Panzer IV were numbered:

801 802
811 812 813 814
821 822 823 824

The 4th company Panzer IV were numbered:

401 402
411 412 413 414
421 422 423 424

They were photographed in March 1941, awaiting shipment from Naples harbour to Tripoli.

Nine Panzer IV tanks from the 8th company arrived in Tripoli harbour on 7 March 1941 and they participated in a military parade in Tripoli.

Above: On the left side of this Panzer IV a battlefield modification has been added. It is a step above the third return roller to help the crew climb up onto the tank more easily.

Left: The rear and side hull armour were only 20mm thick. Although early Panzer IV Ausf.D did not leave the factory with rear turret stowage bins fitted, those that received the 'tropen' hot climate modifications for service in the North African desert did have them fitted.

Above: The two large rectangular hatches on the glacis plate enabled the crew to get access easily to the final drive and braking system.

Right: A smoke candle rack for 5 candles was mounted on the exhaust system. The smoke candle was ejected from the rack by a spring and automatically ignited.

Above: The six and seventh road wheels on the left side have the later cast hubcaps, introduced during Ausf.E production. The only non-standard modification appears to be the addition of the two spare wheel brackets on the sides of the engine compartment, with the resulting move of the cleaning rod brackets to the left superstructure side. The side armour of the Panzer IV could be penetrated by the Soviet 14.5mm anti-tank rifle.

Left: Panzer IV tank nearside idler wheel.

Opposite above: Panzer IV tank suspension assembly.

Opposite below: The round disc is the gun elevation wheel with the azimuth indicator behind it, viewed from inside the right side of the turret. Infront of the wheel is the trigger lamp, trigger circuit wiring, traverse wheel housing and clinometer adjustment.

Left: The handle on the upper left is the power traverse and includes the trigger for the main gun. The motor is for the traverse unit.

Opposite above: Panzer IV tank air cooling system over brake assembly.

Opposite below: Panzer IV tank auxiliary motor for the power traverse gear.

Below: Panzer IV tank driver's seat looking forward.

Left: Panzer IV tank nearside sprocket wheel.

Below: Hull joint method of assembly on the Panzer IV tank.

Right: Panzer IV tank offside track. The track's width was 1 foot 2 inches (356mm).

Below: This German Panzer IV Ausf.D was captured in the desert of North Africa and having been brought back for examination to base camp. Its turret number was 800 and had been assigned to German Africa Korps, 8 Company, Panzer Regiment 5, 5th Light Division later 21st Panzer Division.

Left: Panzer IV tanks had eight double road wheels each side whereas most versions of the Panzer III tank only had six.

Below: Panzer IV tank being driven in the North African Desert. The dust kicked up for the tank was not just a problem for the driver behind it but also meant it made the tank easier to spot form the air.

Knocked-out Panzer IV tank with turret number 813 in the North African Desert.

The armoured elements of the embryonic German Africa Corps in Tripoli in March 1941. Panzer Mk.Is and IIs are lined up on the left and far right, with the Panzer Mk.IVs in the centre.

The same line-up as the previous photograph, Tripoli, March 1941. The jerry cans were to be used to carry water as well as fuel.

Panzer IV tank being driven in the North African Desert.

5

Wire-mesh Skirting on German Tanks

This Panzer IV Ausf.J was assembled at the Nibelungenwerk in September 1943 and used by the Germans to test their Drahtgeflechtschürzen (wire-mesh skirting and its mounting brackets) later that year. After its capture the tank was examined in the First US Army area.

SCHÜRZEN PLATES, METAL SKIRT PANELS

In 1943, Allied military intelligence made errors trying to work out the purpose of the new Schürzen plates, metal skirt armour panels, that started to appear hung on the side of German Panzers. They wrongly believed they were a technological advance system, discovered by German engineers, that would defeat British P.I.A.T and American Bazooka anti-tank shaped-charged projectiles. Instructions were given to Allied infantry armed with these weapons, not to fire at sections of German panzers protected by skirt armour as it was believed they would stop the round penetrating the vehicles armour. In fact, the opposite was true. Shaped-charged projectiles need space to develop the molten hot jet that penetrated tank armour. Trials late in the war found that the use of add-on spaced armour skirts actually increased the performance and penetration of the destructive jet.

Schürzen plates, metal skirt panels attached to the side of German tanks, were designed to stop the armour piercing tungsten carbide core incendiary round fired from Soviet Army 14.5mm anti-tank rifles penetrating the thin side hull armour of the Panzer III, Panzer IV, Panther and Stug III, not the shaped-charged P.I.A.T. and bazooka projectiles. The Soviet 14.5mm anti-tank rifle round could penetrate up to 40mm thick armour. Most German tank hulls had between 30mm to 40mm thick side armour and were vulnerable to these cheap to make, anti-tank rifles.

In 1942, Panzer crews reported that very large numbers of the 14.5mm anti-tank rifle were being issued to Soviet infantry units. They were being trained to fire at the tank's vulnerable side armour and did so effectively. At first, Allied military intelligence was not aware of the damage being caused by these Soviet rifles, hence the misunderstanding.

Initially the German tank designers believed the only way forward to deal with this threat would be to develop tanks with thicker side armour. This is one of the reasons the Panther II prototype tank was developed as it had 60mm thick side armour rather than the Panther Ausf.D's 40mm thick side armour. This was a very heavy resource depleting solution to this problem. A better solution was the introduction of thin metal skirt armour plates. Tests showed that when a captured soviet 14.5mm anti-tank rifle was fired at the side of a Panzer III fitted with skirt armour at a close range of 100m at a right angle, the round broke up and did not penetrate the tank's 30mm side armour. Orders were given that vehicles in production would be fitted with Schürzen skirt armour plates and back fitted to tanks and some self-propelled guns that were already in service.

DRAHTGEFLECHTSCHÜRZEN (WIRE-MESH SKIRTING)

Drahtgeflechtschürzen (wire-mesh skirting) was introduced in September 1944. The wire mesh on the side panels used 6mm diameter wire which gave a 12mm wide centre. Because of the danger of hand thrown enemy grenades being thrown over the side panels and getting stuck between the side mesh panels and the tank hull, that space was covered by long thin panels. The wire used for those panels was sightly thinner. It had a 4mm diameter and the grid had 13mm wide centres.

There were a number of reasons for the change from using solid metal Schürzen plates to Drahtgeflechtschürzen wire-mesh skirting. They were certainly not easier to manufacture. Producing a wire mesh is harder than just cutting out a plate of metal so there had to be some worthwhile advantages to their implementation. One advantage was that they saved weight. Additional weight added to a tank reduced its performance as the engine had to work harder and caused additional strain on the engine, transmission, and suspension.

A more tactical and operational reason for their introduction was that they reduced the amount of dust produced by the tank. A dust cloud was built up between the Schürzen solid metal plates and the tank hull and it was violently expunged upwards and to the rear. Dust clouds produced by advancing tanks could be spotted at great distance by enemy forward artillery spotters. They would issue a fire order and that would result in a lethal hail of high explosive shells on to the tanks position. Anything that reduced the amount of dust produced by tanks was welcomed. Trials showed that the dust build-up between the side plates, tank hull and the track covers were reduced by the fitting of the open weave Drahtgeflechtschürzen. The dust escaped through the holes in the side metal grid rather than being forced, under pressure, out the sides and rear of the solid Schürzen plates.

The solid metal Schürzen plates obstructed the view of the crew from different positions in several directions. They could not see clearly if there was a rock or tree trunk causing an obstruction or if an enemy infantry man had managed to stealthily get close to the side of the tank. The open wire mesh of the Drahtgeflechtschürzen plates enabled the crew to look through the mesh. The open nature of the metal weave enabled the crew to poke branches and foliage through the grid when it was felt necessary to add camouflage to the tanks, so it blended in with its environment.

Protection with the Drahtgeflechtschürzen was similar to the full-plate versions of Schürzen. Firing trials conducted proved that the 5mm plate and 6mm diameter mesh had the same characteristics in absorbing the energy of the Soviet anti-tank rifle's 14.5 mm armour piercing rounds before they hit the tank's side armour.

A Panzer IV Ausf.J was captured and examined in the First US Army area. It was fitted with Drahtgeflechtschürzen wire-mesh skirting. Although the examination report APO 887 was American in origin, a copy was found filed in the National Archives at Kew. This is the transcript.

RESTRICTED

HEADQUARTERS COMMUNICATIONS ZONE, EUROPEAN THEATRE OF OPERATIONS USA (ETOUSA), OFFICE OF THE CHIEF ORDNANCE OFFICER. APO887 27 JANUARY 1945.

ETO ORDNANCE TECHNICAL INTELLIGENCE REPORT NO.131

SUBJECT: WIRE MESH SKIRTING ON GERMAN TANK

Observations by Capt. G. D. Dury, OTI Team No.1 and Lt. D. M. Gilles, Ord. Tech., Intell. Sec., Ord. Serv., HQ., Com. Z., ETOUSA.

A Pz.Kpfw IV German tank equipped with wire-mesh skirting was examined in the First US army area. The skirting was made of 7/32" diameter wire interwoven to form a mesh of approximately 17/32". It consists of six sections, 3 to each side, of the tank, and extends from the top of the superstructure to the bogey wheels and from front to rear of the tank.

The skirting is mounted vertically in line against the outer edges of the track fender guards and is attached to a tubular support by means of brackets bolted to the top of the sections. Small steel straps hold it to the track fender guards.

Although only one tank equipped with skirting was encountered, several damage screens were found discarded along the roads. These appear to be damaged by icy-road accidents and not by weapons.

SIGNED

for the Chief Ordnance Officer
H. N. Toftoy
Col. Ord. Dept
Assistant

Close up of the wire-mesh skirting mounting bracket.

Right: Section of wire-mesh skirting. Note the mounting brackets at the top and steel straps for attaching to the track fender guard.

Below: Drahtgeflechtschürzen wire-mesh skirting was often ripped off the tank as it drove along narrow roads and through woodland.

This Panzer IV Ausf.J at the Musée des Blindés, Saumur (The French Museum of Armour) is fitted with Drahtgeflechtschürzen wire-mesh skirting. It was recovered from a live firing range and restored by the Cadman Brothers in the UK.

Notice the Drahtgeflechtschürzen wire-mesh skirting on this Panzer IV Ausf.J at the Musée des Blindés, Saumur (The French Museum of Armour) covers all the exposed tank side hull armour.

This restored Panzer IV Ausf.J is on display at the Australian Armour and Artillery Museum in Cairns, Northern Queensland, Australia. It has Drahtgeflechtschürzen wire-mesh skirting on the side of the turret as well as the hull.

6

The E 100 (Entwicklung 100) Super-heavy Tank

The E 100 tank was a German super-heavy tank design, the largest model from the experimental development 'Entwicklung' series, which aimed to enhance German armoured super-heavy vehicle production by standardizing simpler, more cost-effective designs. By the conclusion of the Second World War, the E 100 prototype's hull was partially assembled at a workshop at Haustenbeck Henschel tank proving ground at Sennelager near Paderborn in Germany. Work had not started on the turret. The hull was seized by Allied troops and transported to the UK for testing at the Fighting Vehicle Proving Establishment based in Chertsey. The British had intended to complete the vehicle and use it for trials to assess super-heavy tanks' mobility and logistical

This is the Krupp designed E 100 (Entwicklung 100) super-heavy tank hull. It was moved from the small workshop it was housed in to the outside after being captured by the Allies at the end of the War. The drive sprocket outer rings are missing. The turret was designed but not constructed.

expenses for future generations of tanks, but this was halted in favour of the Tortoise tanks, which could do the same thing. The E 100 remained at Chertsey until the 1950s when it was dismantled for scrap, as it offered no significant insights into future tank manufacturing.

Krupp designed the hull of the E 100, but Panzer Kommissioner Ernst Kniekampf believed that Krupp was overburdened with work so gave the job of constructing the E 100 prototype hull to the firm Alder. On 30 June 1943, the engineering staff of Alder started work finishing the design of the E 100 hull and the component parts. It was assembled at the Haustenbeck Henschel tank proving ground in a small workshop building that was just big enough to accommodate this large tank hull. As a result of Allied bombing and scarcity of raw materials, delays occurred in producing parts and transporting them to Haustenbeck. One report stated that only three employees from Alder were assigned to the construction project which also further delayed assembly.

The production vehicles of the E 100 were going to be powered by a Maybach HL 230 P30 V-12 water cooled 700hp petrol engine. One was fitted to the E 100 prototype hull, along with a Maybach OG 40 12 16 B Schaltgetriebe transmission and a Henschel L 801 Zweiradien Lenkgetriebe two radius steering unit. The E 100 had a predicted maximum road speed of 23km/h (14.29mph). For comparison, the maximum road speed of the Maus super-heavy tank was only 20km/h (12.42mph).

The upper hull front armour was angled at 60° and 200mm thick. The lower hull front armour was angled at 52° and 150mm thick. The hull side armour was vertical and 120mm thick with the upper half covered by the removable side skirts that were 75mm thick. The rear armour was angled at 30° and 150mm thick. The hull deck was 40mm thick. The front section of the hull belly plate was 80mm thick and the rear section 40mm thick.

By 17 May 1944, turret drawings had been completed for the E 100. Krupp designed the turret. The turret would have a sloped front and be armed with a 128mm Kw.K. L/55 main gun with a 75mm Kw.K. L/24 gun mounted above it. The design was very similar to the Maus II turret, but it had thinner armour; the front armour was going to be angled at 30° and 200mm thick, the sides were to be angled at 29° and 80mm thick, the 150mm-thick rear armour was to be angled at 15° and the flat roof was to be 40mm thick. The predicted weight of the E 100 fitted with its turret was going to be around 130 tonnes. (The combat loaded weight of the Maus was 188 tonnes).

Contrary to what some people believe, no German vengeance V weapon or single tank design was going to deliver a victory for Germany in the Second

World War. Although you can admire some of the technical achievements developed to make such a large and heavy vehicle, it should be remembered that the German army was forced to abandon Panther, Tiger I and Tiger II tanks if they broke down because they did not have the necessary vehicles to enable them to be recovered—and those vehicles were a lot lighter than the E 100. Germany was suffering crippling fuel shortages. Adding a new, bigger, and thirstier tank to the combat fleet was not practical. The money, time and resources that were necessary to produce the E 100 would have been better deployed in producing more Panzer IV tanks. The E 100 was merely a distraction to the general German war effort. What follows is a transcript of an interview with a German engineer who worked on the E-series of experimental development vehicles, and then a technical team report and a translation of a captured German progress report on the development of the E 100 tank.

CONFIDENTIAL

INTERROGATION REPORT ON GERMAN ENGINEER AND ARMY OFFICER HEINRICH ERNST KNIEPKAMP—SUBJECT E-SERIES TANKS. 31 AUGUST 1945

Question: 'What was the reason for embarking on the design of the E series tanks and what were the general design considerations?'

Answer: 'The reasons were: (1) That to achieve a very strong front plate all possible weight had to be moved to the rear of the tank. (2) Unit motor and power train was desired because of the simplified manufacturing and service. (3) It was planned, by using two types of engines and power train, to produce four vehicles. These four vehicles were as follows: E-10 weighing from 12-15 metric tons, E-25 weighing from 25-30 metric tons, E-50 weighing from 50-65 metric tons, E 100 weighing from 100-130 metric tons. This program was first conceived by Herr Kniepkamp in May 1942 and was submitted and approved as a project in April 1943. (4) All suspensions should be attached from the outside and no fighting space should be encumbered by through torsion rods. (5) In case the front idler or any of the bogie wheels were destroyed by mines, the vehicle must be capable of proceeding by adjusting the track around the remaining wheels. The technical superiority of forward drive was recognized, but the military advantage of having the drive at the rear where it was not endangered by mines or gunfire influenced the choice of rear drive. The rear drive was also favoured because of the clean fighting compartment achieved thereby.'

Question: 'What were the considerations relating to the location of the centre of gravity of the vehicles with respect to the ground contact of the track and the centre line of the turret?'

Answer: 'In general, it was desirable to have the centre line of the turret, the centre of gravity of the vehicle, and the middle point of the ground contact as close together as possible. In practice the centre of gravity of the vehicles, for example the Panther, tended to be forward of the centre of track contact and a slight forward heaviness was considered desirable.'

Question: 'What were the most important features of the E-10?'

Answer: 'One of the features of the suspension of the E-10 is the fact that the suspension wheels can be raised allowing the tank to sit on its bottom. Since the ground clearance is 450mm, this means that the tank can be lowered this amount. The mechanism for this feature is a simple spindle with a power take-off drive from the main motor. When the vehicle is operating on smooth ground it is possible to lower the suspension to any extent desired. The weight of the lowering mechanism is only 250kg.'

Question: 'Were the drawings completed?'

Answer: 'In the summer of 1944 the drawings were completed and an order given to manufacture three pilot models. The order was given to Klockner-Humboldt-Deutz, Ulm. The drawings were believed to be in the KHD Werke (factory) No.1 in the town of Ulm. It was known that the drawing office had been destroyed but Herr Kniepkamp believed that the drawings had been saved. They could easily be redrawn by the firm at a cost of a few thousand marks.'

HISTORY AND GENERAL CHARACTERISTICS OF THE E-10

The designer was Herr Honsig. The proposed motor was HL-101, 12-litre, giving 500hp at 3,800rpm. In the prototype the HL-90 10-litre was contemplated. Both motors were conventional 12-cylinder V-types. HL-101 employed solid injection. HL-90 was equipped with a carburettor. There were three possible transmissions: (1) Zahnradfabrik Electro-magnetic, (2) Maybach-Olvar 8-forward and 4-reverse speeds. (3) Voith torque converter. Steering mechanism: (1) Mechanism similar to Tiger II (2) Steering mechanism with the substitution of two Voith torque converters for the four conventional clutches in the Tiger II steering mechanism. (3) Positive displacement axial piston pump and motors built by Zahnraderfabrik, Augsburg.

There were three pilot models under construction. The three hulls were made in Silesia but were not finished when the Russians got there. Proposed weapon:

armament was to be a six-barrel 7.5cm rocket projector or a 7.5cm gun L48 in limited traverse mount. The crew consisted of three men: driver, gunner and loader. A stabilized line of sight was planned for the gun. Armour; 20mm sides; 50mm front, approximately 60° from vertical. Running gear: four wheels with steel-rimmed resilient wheels, 850-900mm diameter. Track width approximately 40cm. Maximum speed 50km/h.

GENERAL CHARACTERISTICS OF THE E-25

The E-25 was essentially the same as the E-10 except: (1) engine was to be supercharged to produce 700hp, (2) armour: sides 30mm, front 70mm. There were two proposals: one vehicle was to be built with a turret and to be used as a scout vehicle, and the other was to be without a turret and used as a tank destroyer. Armament on both vehicles was to be the 7.5cm gun L70.'

Suspension: four wheels of 1m diameter. The ground clearance was not adjustable. Maximum speed 50km/h. Crew of four men. The design was made by Argus at Karlsruhe, and all activity was evacuated to Heimschule, Hermansberg near Pfullendorf where Dr. Klaue was in charge. (It is known that all drawings were taken from Dr. Klaue by the French early in May 1945). Three hulls were reported to be at Alkett in Berlin-Spandau. (It has since been determined that these hulls were no longer there.)

GENERAL CHARACTERISTICS OF THE E-50 AND THE E-75

The suspension was to use six units per side. The same units were to be used on the E-75, but nine instead of six preside. All the wheels would have the same diameter, 850mm. Instead of having light twin wheels like on the Tiger, single, staggered, resilient wheels were called for in order to get away from cramping. The springs were under construction at Wasserchutte, Oyenhausen. The combat track for the E-50 was to serve as the shipping track for the E-75.'

Two motors were considered: HL234 rated at 900hp at 3,000rpm, with dual-injection. The same motor with supercharging was rated at 1,200hp. If operated as a diesel, which was accomplished by using a different cylinder head, it gave 650hp at 900rpm with supercharging. A third motor which was completely interchangeable with the HL234 was the Klockner-Humbolt-Deutz 8-cylinder, 2-cycle V-type, water cooled, 30-lt diesel rated at 750-900hp at 2,000rpm. Scavenging was handled by a pump that was driven by a small motor.'

'Three transmissions were being considered: (1) Olvar gearbox similar to Tiger II, (2) Voith torque converter, (3) 'Mech-Hydro', being jointly designed by Maybach, Zahnradfabrik, AEG and Voith. The latter transmission was designed as a combined transmission and steering unit in conjunction with a positive displacement hydraulic pump and motor system. The Mech-Hydro' drive was tested for four years in a railway car and was supposed to be installed in the E 100 in May 1945. The motor could be rated at 10% above normal by using this transmission because of the hydro-dynamic element in the transmission.'

Question: 'What were the hulls on the E-50 and E-75 to be like?'

Answer: 'The hulls on the E-50 and E-75 were to be similar to Panther and Tiger II except for thicker front armour. The tracks were approximately the same as the present Tiger and Panther tanks. In all this work nothing had been completed. Development of components was in progress. No plans for complete vehicle assembly had been formulated.'

GENERAL CHARACTERISTICS OF THE E 100

The E 100 was started with a rear drive but only on paper. The front drive version was reasonably well on the way and was intended to be ready and assembled by the summer of 1944, but bombing had disrupted the program. The design with rear drive was barely started.'

Question: 'How many suspensions for the E 100?'

Answer: 'There were only two; the Jenschke design with twin dual coil springs and the Auto Union design as in the other 'E' series tanks.

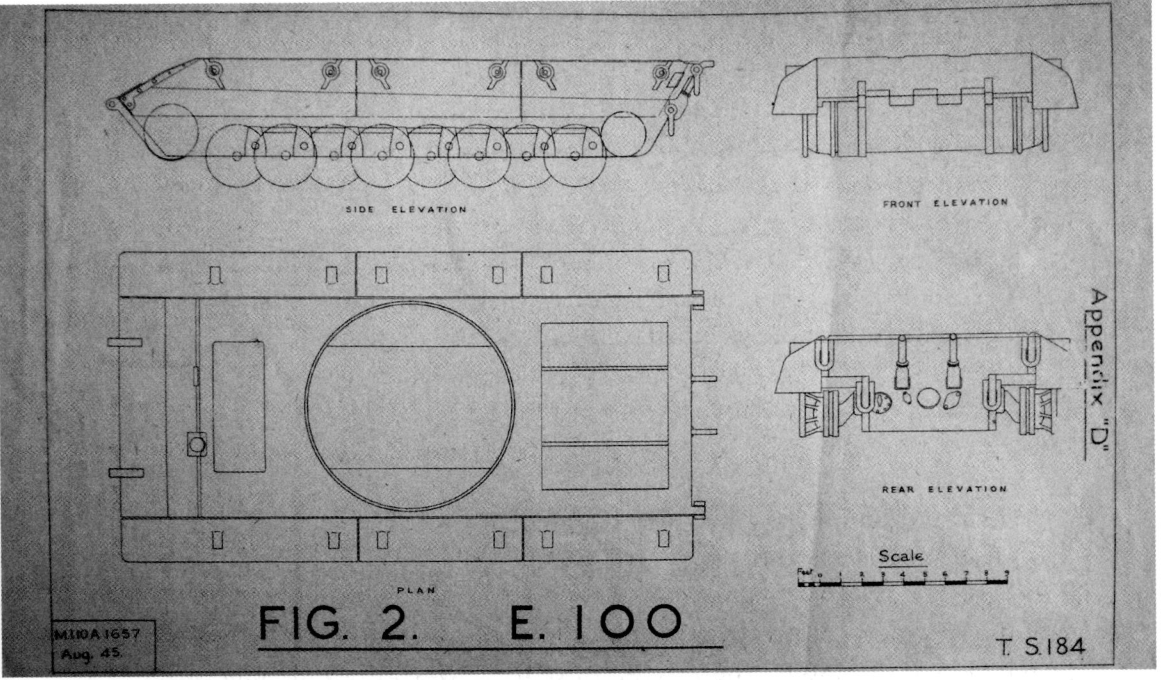

British War Office Ministry of Intelligence section 10 (M.I.10) technical drawing of the E 100 hull.

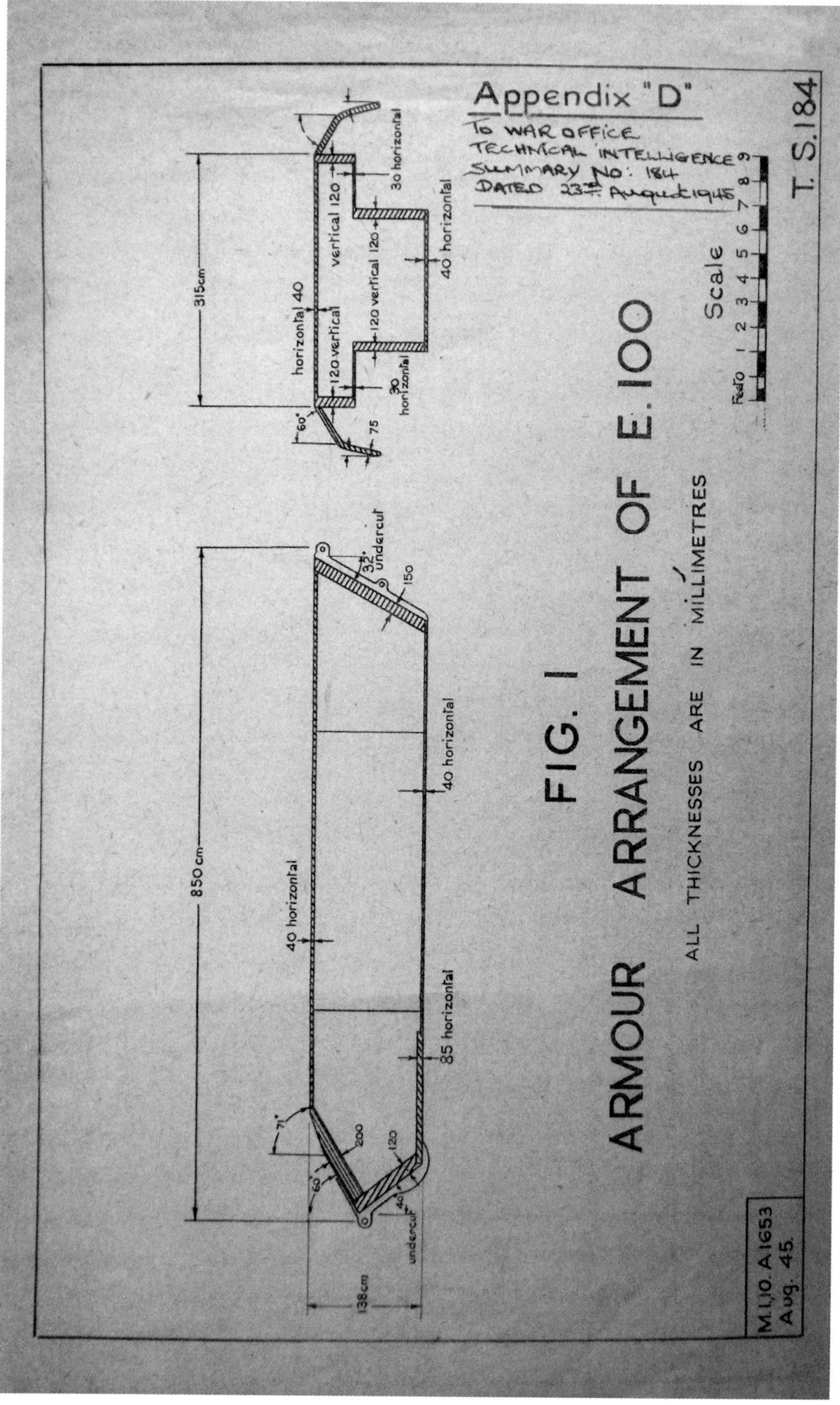

British War Office Ministry of Intelligence section 10 (M.I.10) technical drawing of the armour thickness and arrangement of the E 100. Thickness in millimetres unless stated otherwise.

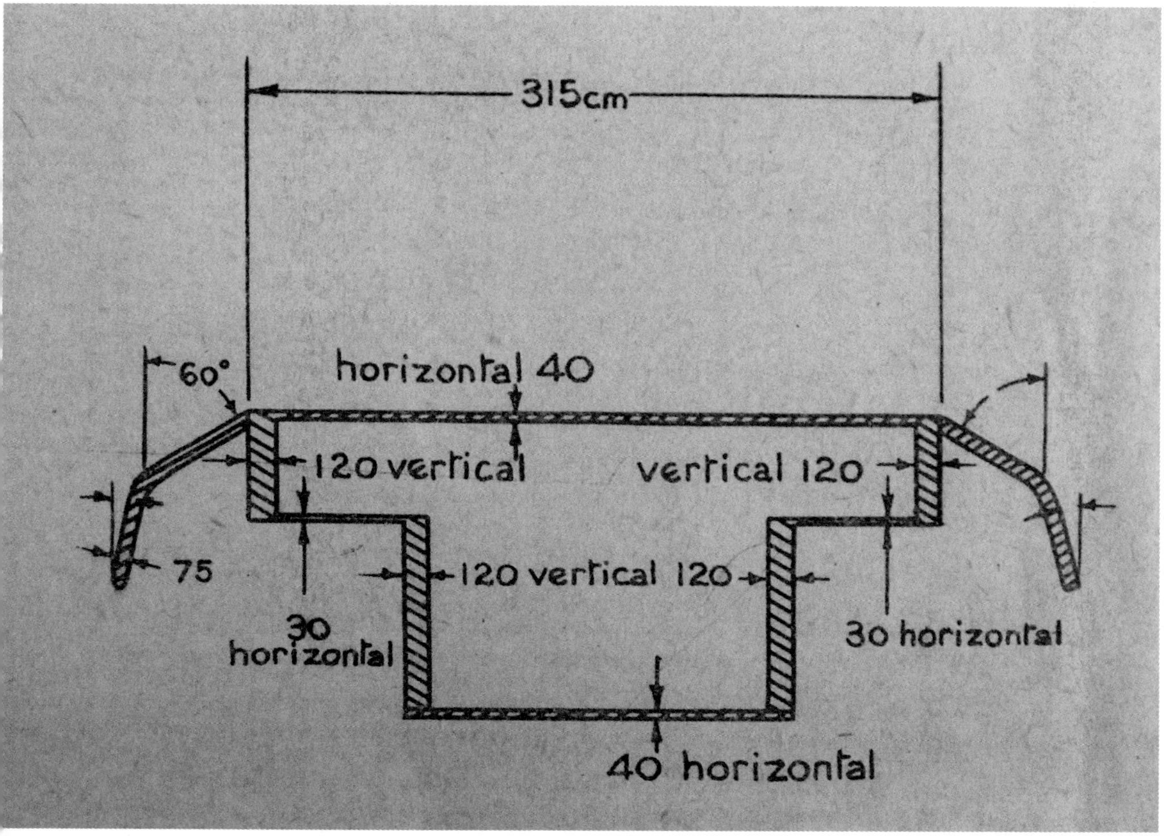

British War Office Ministry of Intelligence section 10 (M.I.10) cross-sectional view technical drawing of the armour thickness and arrangement of the E 100. Thickness in millimetres unless stated otherwise.

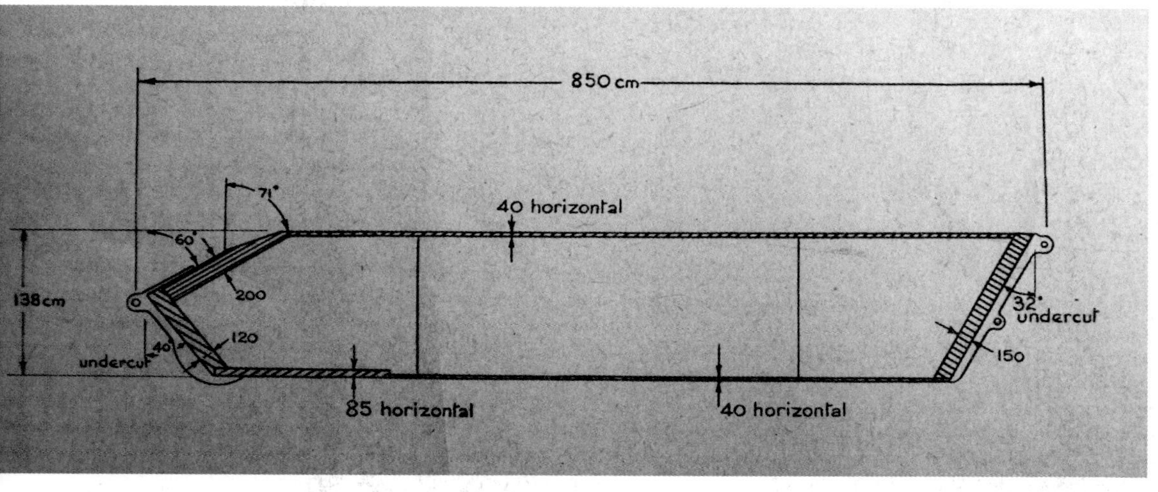

British War Office Ministry of Intelligence section 10 (M.I.10) side view technical drawing of the armour thickness and arrangement of the E 100. Thickness in millimetres unless stated otherwise.

RESTRICTED

HEADQUARTERS COMMUNICATION ZONE, ETOUSA, OFFICE OF THE CHIEF ORDNANCE OFFICER
APO 887. 24 MAY 1945.

ETO ORDNANCE TECHNICAL INTELLIGENCE REPORT NO 288

SUBJECT: PRELIMINARY REPORT ON HENSCHEL TANK PROVING GROUND

GENERAL

The Henschel tank proving ground and development centre near Sennerlager, Germany, was given a preliminary examination. Of greatest interest were an experimental heavy tank and an experimental gun motor carriage found in the final stages of assembly. Special projects were developed at the proving ground, including waterproofing track studies and other subjects allied with armoured fighting vehicles, the scope of which could be ascertained from the files of recovered documents. The immediate area is built to resemble a rural farm, it included administration buildings, two separate workshops, and a specially constructed concrete pool for waterproofing test. Material located at the proving ground is as follows:

a. All documents available at the time were moved to Ordnance Section, Ninth U.S. Army.

b. One tank, model No.100E, of approximately one hundred and ten tons, under construction.

c. One self-propelled gun chassis, under construction, apparently for 17cm gun. Believed to be known as the Grille 'Cricket'.

d. One Jagdtiger 12.8cm gun, in running condition.

e. Two ploughs, experimental, use unknown.

f. Special concrete pool for waterproofing projects and submersion tests.

g. Diving suite for waterproofing tests.

h. Two early types of experimental tanks.

i. One Tiger Model E test vehicle.

j. One Tiger Model B test vehicle, old style turret.

k. One Russian tank, demolished.

l. Six types of tank track.

m. Four tank engines.

n. Two tank sights; TFZ 9b/l, from Tiger Model B and WZF 2/1 blc from Jagdtiger.

As well as discovering the E 100 hull, Allied troops also found the super-heavy German experimental self-propelled gun called Grille 'Cricket' at the Haustenbeck proving ground at Sennelager near Paderborn.

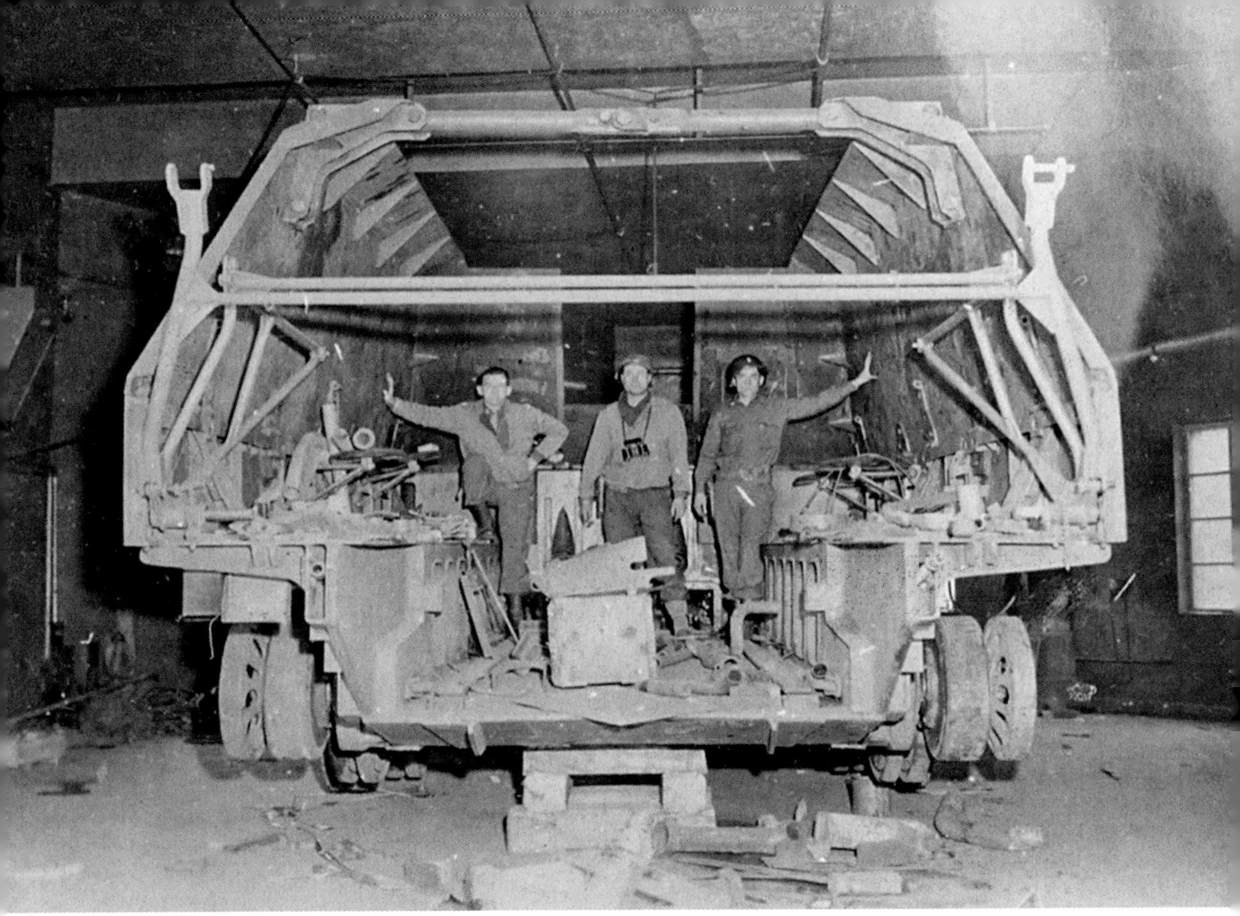

Rear sectional view of the super-heavy German experimental self-propelled gun called Grille 'Cricket' showing the interior and the track from the gun carriage.

Upper carriage of the 21cm gun Mrs.18 chassis and 17.2cm K72 L/50 gun tube for the German experimental super-heavy German experimental self-propelled gun called Grille 'Cricket'.

The muzzle brake for the 17.2cm K72 L/50 gun that was going to be fitted to the super-heavy German experimental self-propelled gun called Grille 'Cricket'.

Concrete pool for watertight and submersion testing at the Henschel tank proving ground and development centre near Sennerlager, Germany.

End view of the armoured track guards for the E 100 at the Henschel tank proving ground and development centre near Sennerlager, Germany.

Side view of the armoured track guards for the E 100 at the Henschel tank proving ground and development centre near Sennerlager, Germany.

Inner side of the front section of the armoured track guards for the E 100 at the Henschel tank proving ground and development centre near Sennerlager, Germany.

Left side of the E 100 tank at the Henschel tank proving ground and development centre near Sennerlager, Germany.

Left side of the E 100 tank, showing supports and threaded sections for spaced armour at the Henschel tank proving ground and development centre near Sennerlager, Germany.

Top view of the E 100 tank showing turret opening at the Henschel tank proving ground and development centre near Sennerlager, Germany.

Combat tracks for the E 100 tank. The width of the track was 39 9/16 inches (100.48cm).

Rear view of the E 100 tank hull.

Composite side view of experimental heavy armoured self-propelled artillery chassis of the 17cm Grille 'Cricket'.

The E 100 hull being examined by Allied troops, 20 April 1945.

ITEM NO.18

FILE NO. XXVIII-3

RESTRICTED

TRANSLATION OF GERMAN PROGRESS REPORT ON DEVELOPMENT OF THE E 100 TANK.

COMBINED INTELLIGENCE OBJECTIVES SUB-COMMITTEE.

TABLE OF CONTENTS

Subject

PROGRESS REPORT NO.291 ON VEHICLE E 100

Location: Haustenbeck i.L. Date: 15 January 1945 Type: E 100

In accordance with the travel report: —In accordance with main group book—In order to give a general overall picture of the work done on E 100 up to the present date, sixty-four different pictures have been made for this purpose. Every photograph has an explanation attached to it. Because of the small building in which the vehicle was located it was impossible to take photographs from all angles. However, it is hoped that the attached photographs will suffice to give a clear picture of how far the work has progressed.

Due to many difficulties caused by war conditions, it was not always possible to receive missing parts at the time they were needed, therefore the assembly of the vehicle was slowed down. It should be pointed out that only three employees of the Adler plant are available for this work. The missing coil springs which are necessary for the suspension assembly have been mis-sent due to some mix-up. The three Adler employees informed me that the assembly of the suspension can be completed as soon as the missing coil springs arrive.

In regard to the track, we have only received the transport track, the combat track has not yet arrived. Parts for the assembly, which lies between the bulk-head and the rear wall, are on hand except for the missing fuel leads. The parts which were needed for the fighting compartment and inside apparatus have arrived and are being installed. Further photos will be made after the assembly has been completed. The drive of the sprocket is accomplished through the final drive with brakes, steering mechanism, change speed gear box, and propeller shaft from the engine. The cover plate for the main gear box and steering gear has not arrived.

The Electrical Installation will be made after the missing gear cover plate on which the instrument panel is mounted arrives. As soon as the fuel leads arrive and the Electrical Installation has been completed, the transmission can be tried out. Henschel will be told about the delivery of the cover plate. A list will be made of the small missing parts so that they can be delivered in the shortest possible time by the Adler plant.

As to information about the turret. Neither the date of delivery of it, nor the equivalent traveling weight is known: we want to know how the situation stands. This question is brought up so that the equivalent weight can be transported by me from the experimental area to the assembly shop. When the dimensions and the weight are known to me transport can be arranged.

In case something has been forgotten in compiling this progress report, explanations should be asked for. In the following pages are photographs with attached explanations. For better survey they are divided into:

Group 1: hull with accessories
Group 2: engine compartment and radiator and cooling fan compartment (floodable)
Group 3: electrical installations & apparatus
Group 4: fighting compartment with its installations (as far as finished)
Group 5: left suspension assembly
Group 6: right suspension assembly

E 100 viewed from front sideways to rear. Suspension arms visible as well as front drive wheel minus sprockets.

This is the front of E 100 with showing the right drive sprocket wheel minus the sprocket ring. Details of the final drive housing can be seen. Headlamp mounted in the middle of the front glacis plate. The front towing eye was part of the side armour plate. The mounting eye for the track guard was built into the front armour plate.

Rear of E 100 with left and right idler wheel mounted.

Left: This is a view of the left rear side of the E 100. The idler wheels are mounted on crank arms with the track adjusting mechanism mounted inside the hull. The track armoured side guards have not been fitted. A cylindrical convoy taillight is fitted above the idler wheel.

Below: Right lower rear section with track tensioner cover, to the left of it is the cover for the flywheel inertia starter and above that is the armoured protection for the exhaust manifold exit point from the rear armour plate.

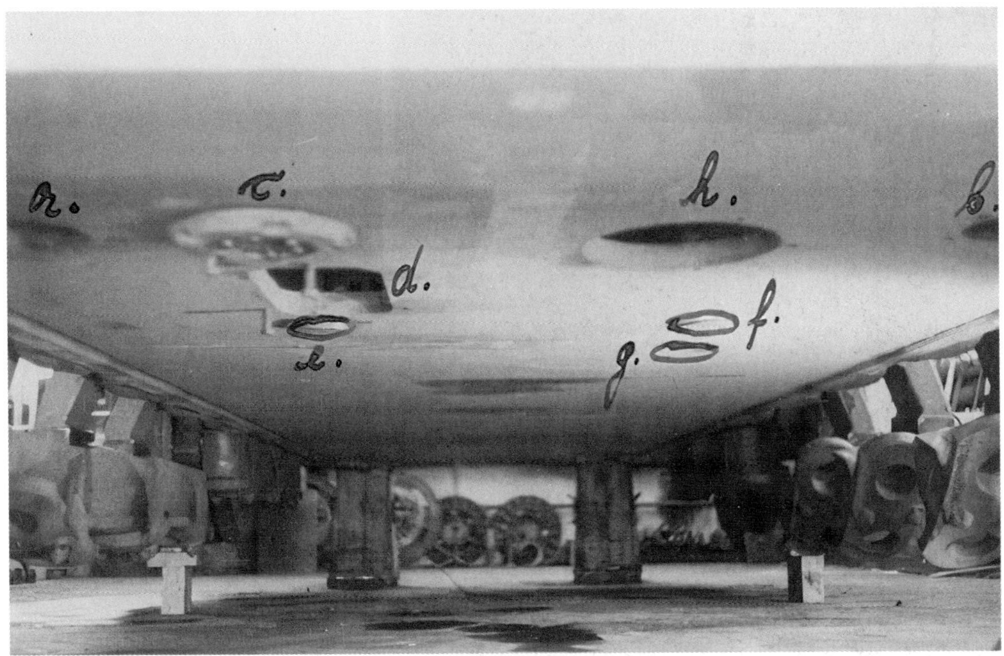

Above: Under-floor of the E 100 with left and right suspension crank arms for bogie wheels and fastened on them the spring seats. Holes (a) and (b), which are closed from bottom with covers, are used for better access to the inside parts of the idler adjusting crank arm. The cover (c) is for the water drain valve of the radiator. The hole (d) is to give better access to the water pipes and to the hose supports for more accessibility, when work is to be done in that part. The holes (e) and (f) are used for taking up the floor flap. In the opening (g) are the plugs for the fuel drain fitting. The hole (h) has the purpose of giving access to the starter.

Right: Engine cover being lifted with left and right grating sections visible, also rear section of the turret ring cutout can be seen.

Left: Engine hatch cover opened.

Opposite above: Engine compartment hatch cover open with (a.) mounting location for the air intake to the ventilator Johnlüfter (John fan), (b.) fuel filling cap and (c.) radiator filling cap. The two air intake grills with wire mesh covers are visible on the left and right. Johnlüfter was a brand name adopted in German technical nomenclature, for a specific type of air ventilator.

Opposite below: Engine compartment top deck cover removed showing top of engine and cooling system installation as well as part of the fighting compartment.

Below: Engine hatch cover closed.

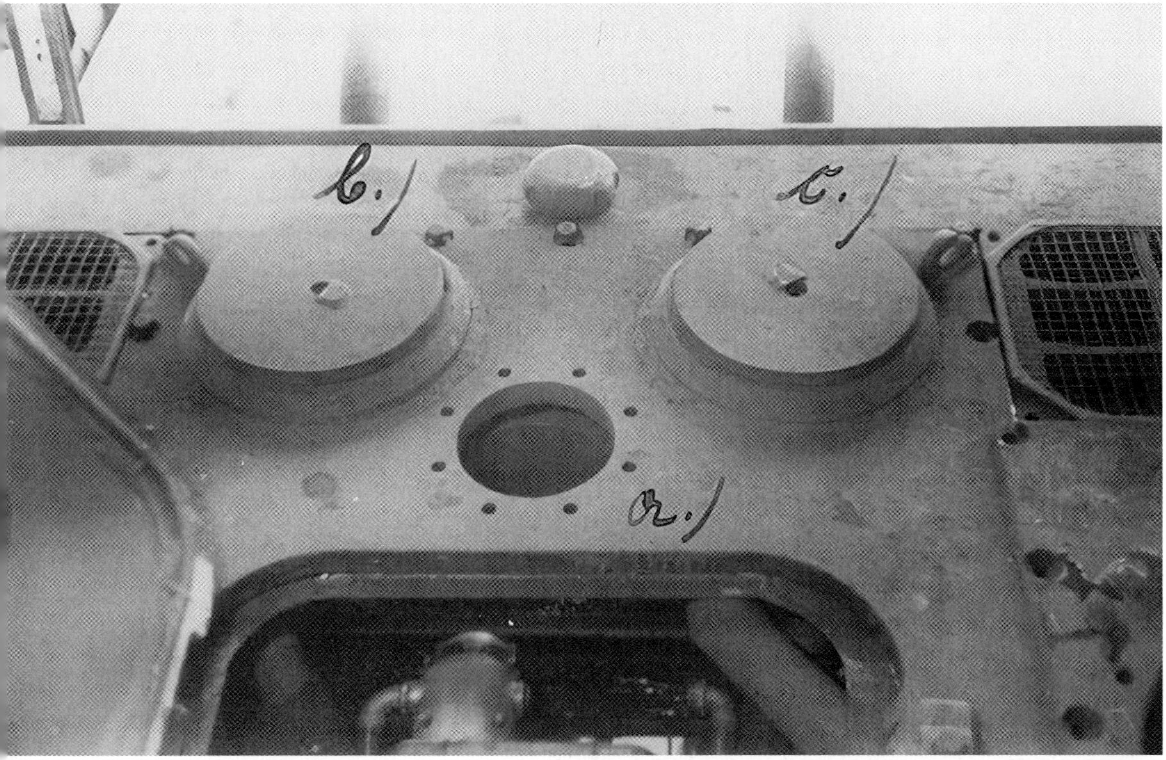

Above: Engine compartment top deck cover removed showing the Johnlüfter (John fan) and cooling system installation.

Left: Right-hand cooling assembly with covered grills and installed cooling fan visible.

Opposite above: Right forward air intake grating with installed wire mesh.

Opposite below: Right cooling assembly cover with cooling fan blades visible.

View (towards rear) from the turret ring of the top plate with its covers and seals etc... cooling fan blades not installed.

Six sections of the E 100's armoured track guards had been delivered but they were found stored outside. They would be mounted over the suspension. The inside view of one of the sections can be seen in this photograph.

Six sections of the E 100's armoured track guards had been delivered but they were found stored outside. Outside view.

Top plate of the hull with opening for removing and installing the main gearbox and steering gear. View looking towards the front. Inside two shock absorbers can be seen located on each side wall. In the middle is the covered transmission and main drive shaft cover.

Above: This is the front of E 100 with showing the left drive sprocket wheel minus the sprocket ring. Details of the final drive housing can be seen. Headlamp mounted in the middle of the front glacis plate. The front towing eye was part of the side armour plate. The mounting eye for the track guard was built into the front armour plate.

Left: View of the right-hand cooling assembly as viewed from the front. On the right side of the picture is the fuel filler cap.

Above: View of the rear of engine compartment. On the left of the picture is the fuel tank filling cap and on the right is the water tank filling cap.

Right: Top view of engine air cleaners. Water and fuel tanks are in the background with filling caps on show.

View of the engine compartment and right and left cooling assemblies as installed.

Connections of the right air intake pipe as well as pipeline of the radiator system. In the middle of the picture is the oil inlet for the right radiator cooling fan.

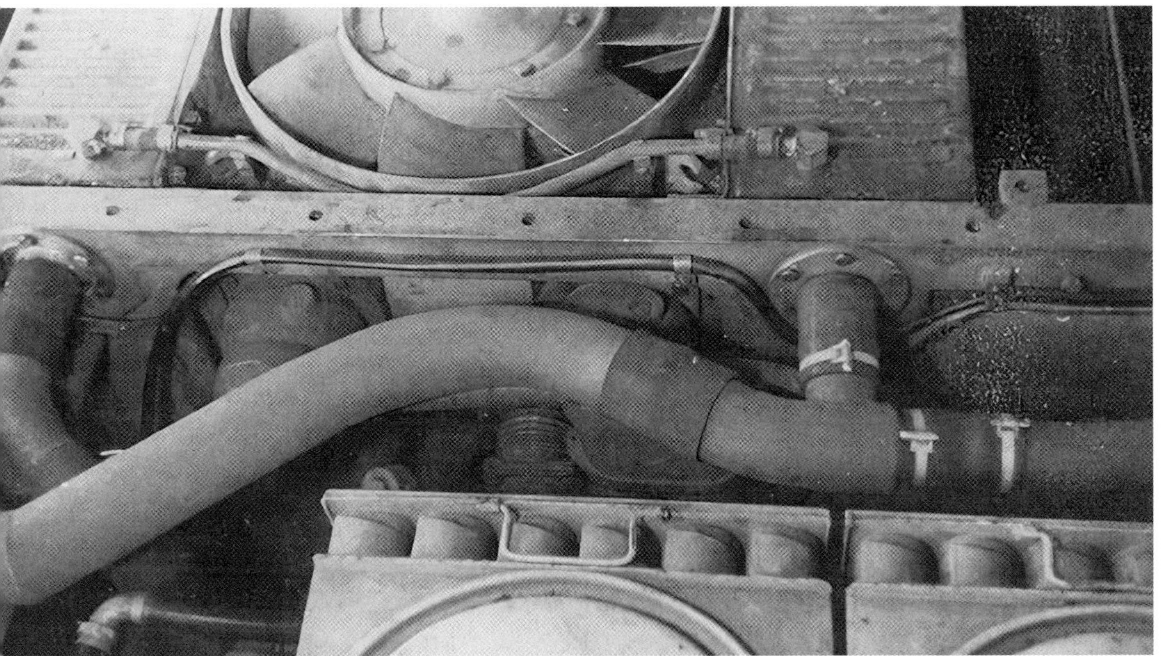

View of the radiator and fan looking across vehicle from right to left with its piping system in the engine compartment visible.

View into the rear part of the engine compartment, between the rear air cleaner and the fuel and water filling tank (the left magneto has been removed).

Above: View from top through the turret ring of the bulk head wall. On the right side of the picture, next to the propeller shaft, immediately on the bulkhead wall, are two parallel cables which lead to the main battery switch. In the middle of the fighting compartment is the fuel tank with recess for the propeller shaft. On the left side of the shaft on the bulkhead wall is the electric automatic fire extinguisher. On the hull floor in front of the bulkhead wall, that is, on the right and left side of the fighting compartment, are both 12-volt battery cases with heating plate installation.

Left: Left battery case with installed heating plate is lying on the floor near the bulkhead wall in the fighting compartment. In front of it are the plugs of the bogie wheel crank arm bearings. Opposite, is the fuel tank in the fighting compartment.

Right: Right battery cage with installed heating plate is lying on the floor near the bulkhead wall, in the fighting compartment. On the right side of the battery is the container of the automatic Minimax fire extinguisher equipment. (The primer pump is on the right, next to the Minimax). Between the right hull wall and the fuel tank, lying in the middle of the floor, is the armoured tube for the cable circuit. It connects the instrument panel, voltage regulators and batteries.

Below: View of bulkhead wall with mounted switch panel with its switches. On the left is the electric automatic inimax. The cables leading down from the selection switch go to the left and right battery.

Above: Switch panel with the following instruments: a) switch protector for self-starter; b) main selection switch; c) anti-interference for control lamp; d) voltage regulator of the generator; e) anti-interference of generator; f) wall socket; g) holes for tie cables leading to the engine room.

Left: View from bulkhead wall toward front of hull interior. On right, next to the main gear box is the armoured tube with its circuit, which is connected to the instrument panel.

Above: View of the fighting compartment seen from the top through the tower opening in the direction of the main gear box. In the middle is the cover for the propeller shaft. At the bottom of the picture is the fuel tank which leads nearly to the bulkhead wall. On the left and right, in the hull, is the top gusset. On the right and left in front are two shock absorbers, two on each side. Curved segments were cut out of the inside surface of the hull sides to give clearance for the turret basket.

Right: View from the turret opening to the bulkhead wall. Bulkhead wall with the fire extinguisher, primer pump, switch board & electrical instruments as well as duct for the cooling of the exhaust jacket with Its roto type valve. On the right and left in front of the bulkhead wall are the battery cases. In the middle of the floor is a fuel tank with a depression in the centre, through which the propeller shaft passes from the motor to the main gear box. In the foreground is the propeller shaft once more with a protector, which goes as far as the main gear box.

Left: The picture shows clearly the roto-type valve of the duct for the admission of cooling air for the transmission and its exhaust jacket. The ends of the two ducts go to the bulkhead wall.

Opposite above: View from the left fighting compartment wall up to the wireless operator's seat perpendicular to the propeller shaft.

Opposite below: View from the bulkhead wall into the inside of the hull. On the right and left are two shock absorbers, two on each side, one behind the other. In the middle is the propeller shaft protector and the sheet metal cover of the gear box with the duct on left, next to the propeller shaft. On the left side of the main gear box is the driver's seat raised to the high position and the steering wheel lying in front of it with the Argus steering glide cage mounted on the steering gear. The preselection gear lever for the 8-speed semi-automatic transmission was on the driver's right.

Below: View from the right fighting compartment wall up to the driving seat perpendicular to the propeller shaft.

Above left: Photograph from the right bulkhead wall in the direction of the wireless operator's seat. On the right foreground is the flange of the final drive shaft. Behind are the shock absorbers, on the side wall.

Above right: Second shock absorber, on left side with its top & bottom connections.

Locking cover of the bogie wheel crank arm installation seen from the inside of the fighting compartment. The pipe on the floor is longitudinal direction and leads from the bulkhead wall to the Instrument panel on right of the main gear box.

Above: This is the propeller shaft slip joint leading through the bulkhead wall and flanged onto the engine. On the right, next to it, are the two cables leading from the main breaker switch to the batteries. Further on the right is the roto-type valve for the cooling of the exhaust pipe jacket. Under the propeller shaft (in the fighting compartment) is the fuel tank.

Right: View from bulkhead towards gearbox showing intermediate bearing partly covered and propeller shaft joint universal.

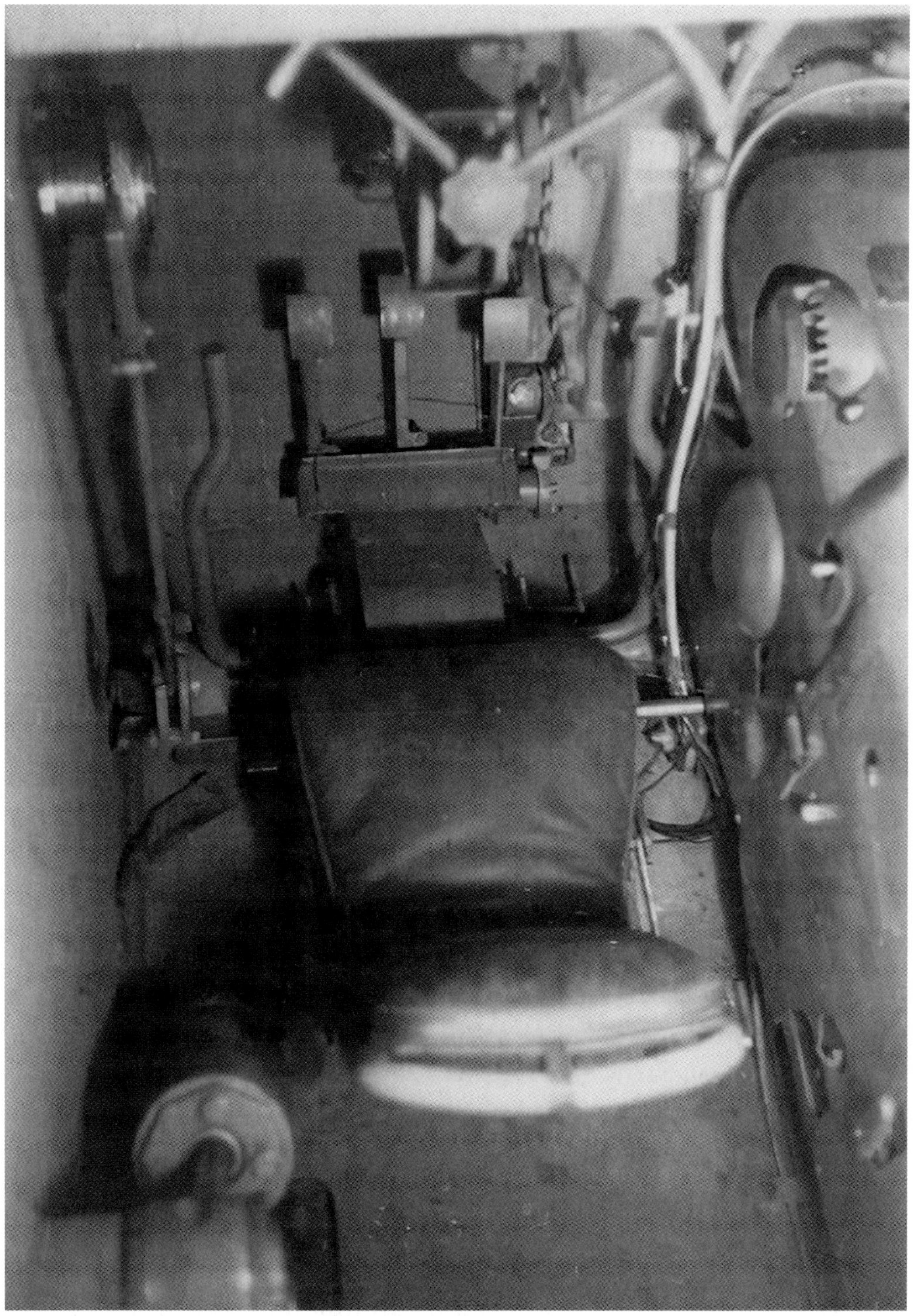

Driver's seat with steering wheel and pedal arrangement. On the left is the first shock absorber with bearings. The preselection gear lever for the 8-speed semi-automatic transmission was on the driver's right.

Driver's seat in down position. First shock absorber behind it.

Driver's seat elevated to high position with its fastening to the floor and the steering wheel, front, as well as the opening for the emplacement of the driver's periscope.

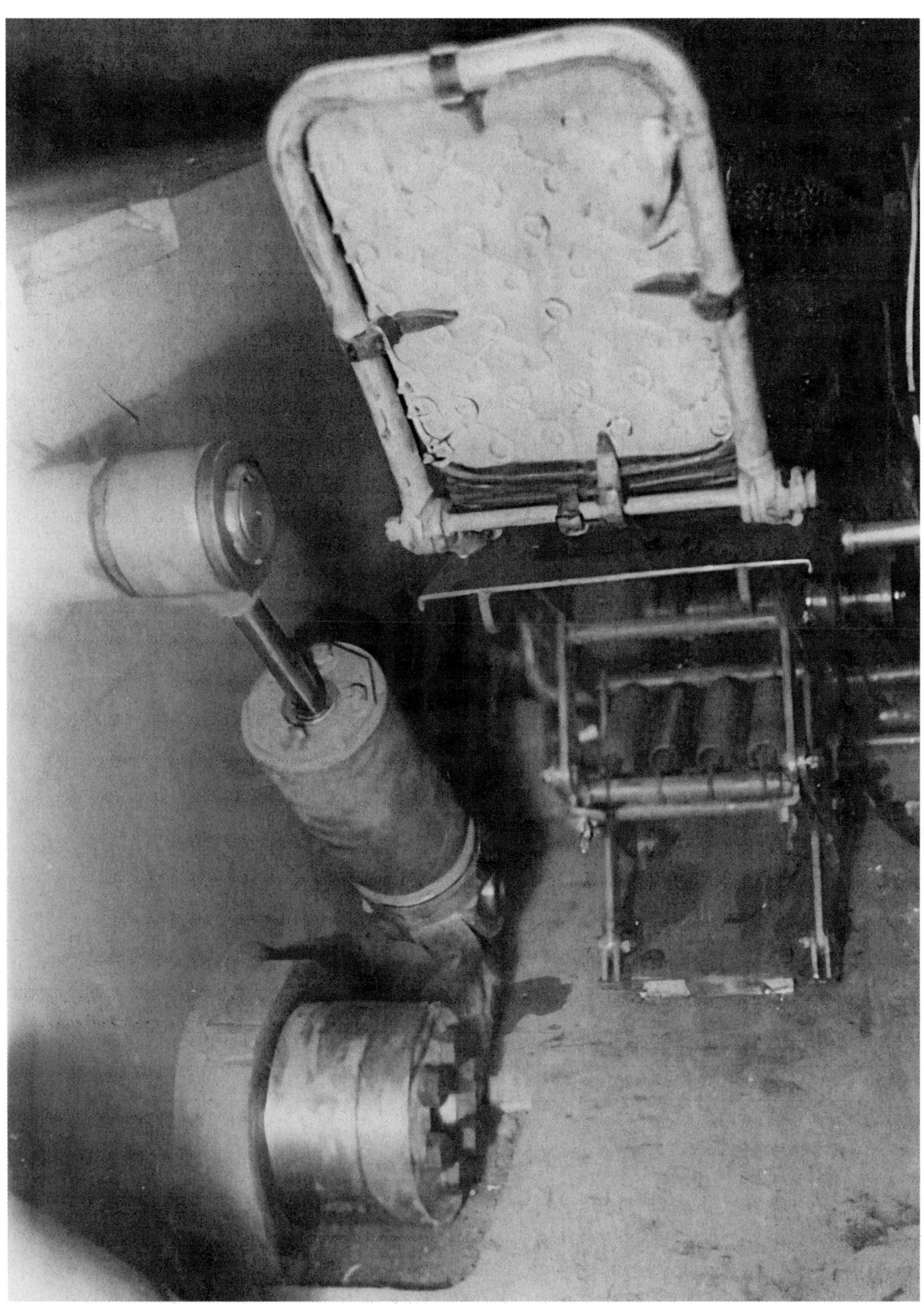

Driver's seat elevated.

View from the left bulkhead towards the driver's seat with visible main gear box and steering gear (rear).

Left side of suspension with mounted final drive sprocket hub, idler, part of cranks with and part without bogie wheel as well as supports for the helical spring and spring anchors.

Left side of suspension. The view is from front to rear.

Left: The front of the E 100 showing the left drive sprocket wheel minus the sprocket ring. Details of the final drive housing can be seen. Headlamp mounted in the middle of the front glacis plate. The front towing eye was part of the side armour plate. The mounting eye for the armoured track guard was built into the front armour plate. On the hull wall over the first road-wheel can be seen an attaching point for the armoured track guard.

Opposite above: Sprocket of sprocket hub and first bogie wheel of the left suspension assembly.

Opposite below: Road wheel hub plate and the spring plate fixed to crank arm.

Below: Sprocket hub of the left road wheel side without sprocket ring (oblique rear view). Details of the final drive housing can be seen.

Opposite above: Left suspension assembly with sprocket hub front bogie wheels, and with bogie wheels not yet mounted on the crank arm. Above are the upper spring supports and bumper stops.

Opposite below: Left suspension assembly with staggered set of bogie wheels.

Right: Right suspension seen from the sprocket hub in the direction of the adjustable idler wheel crank arm. On the hull side walls are the attaching points for the armoured track guards.

Below: Left and right suspension crank arms with bogie wheel hubs as seen from under the hull floor.

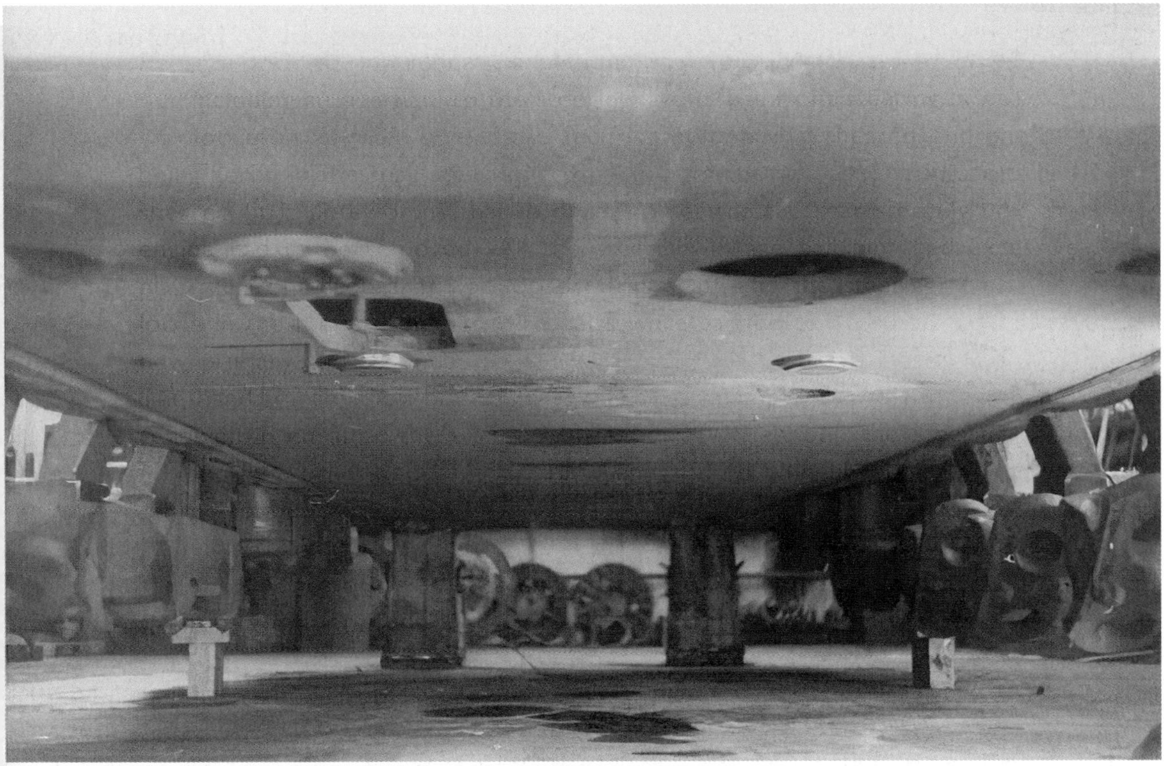

Right suspension crank arms with mounted spring supporting plates (taken on from bottom).

Right idler wheel crank with mounted idler. View in direction of sprocket.

View of the right side and rear of the hull. The curved armoured exhaust protective covers can be seen attached to the rear of the hull.

Right side of suspension. View from the idler in the direction to the sprocket with some mounted road wheels.

Left: The E 100 on a transportation trailer 5 January 1946. The front sprocket wheel is visible, minus the sprocket teeth ring. Above it the thickness of armour on the front glacis plate can be seen.

Opposite: British soldier inspecting the captured E 100 tank hull outside the workshop.

Below: The E 100 on a transportation trailer, 5 January 1946.

EDITOR'S NOTE

In a large update report on the progress of 1946 British Cold War tank design projects stored in the National Archives at Kew there was a small paragraph about the captured German E 100 tank and a line drawing.

FIGHTING VEHICLES DIVISIONAL LETTER NO.15

E 100–GERMAN TANKS

It is proposed to re-build the German E 100 for trial purposes, but the project has been postponed owing to its low priority and the pressure of other work. The accompanying sketch shows the tank reconstructed from drawings.

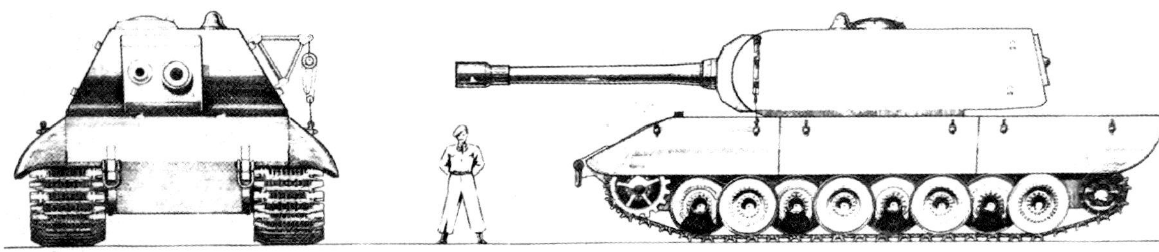

British pictorial construction of the German E 100.

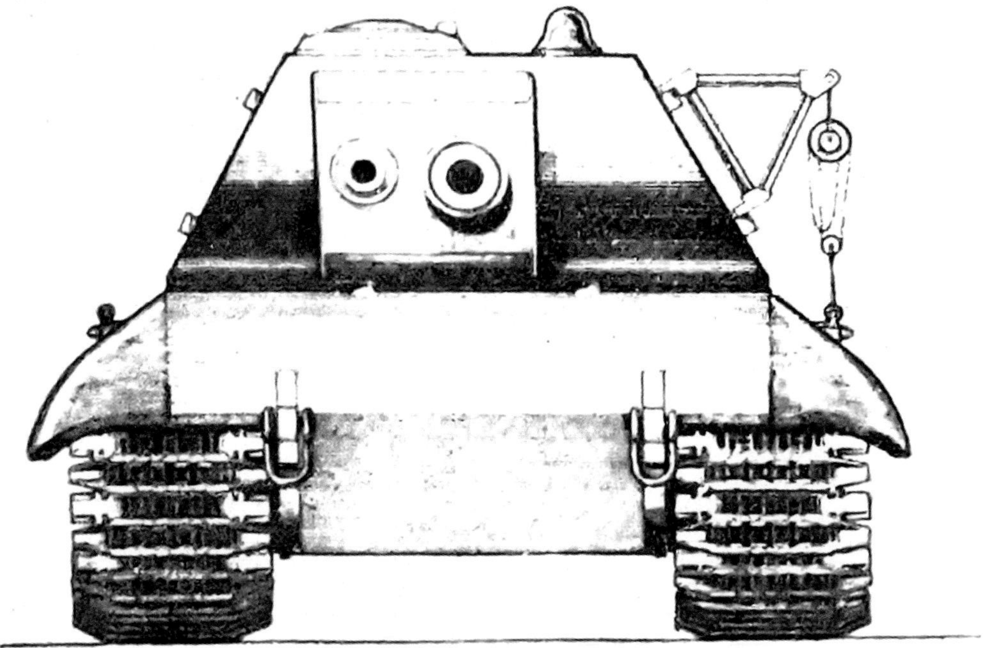

British pictorial construction of the front of the German E 100 showing the turret in position and the ammunition loading crane on the back.

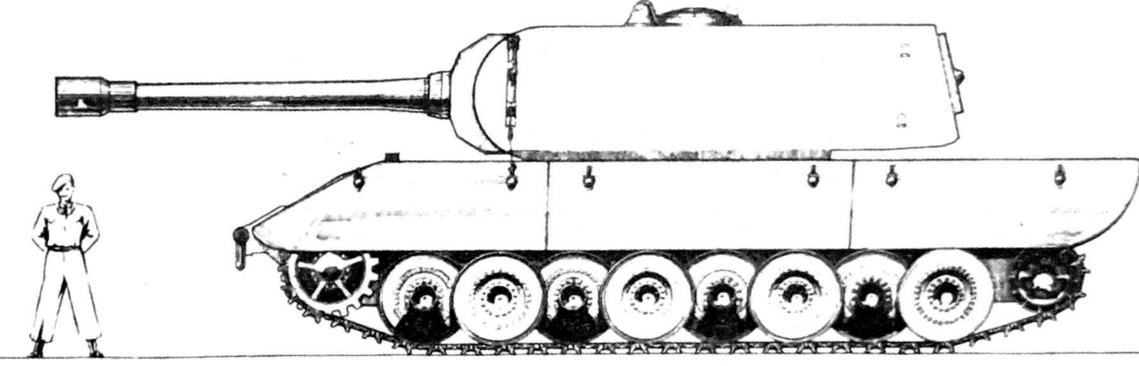

British pictorial construction of the side of the German E 100.

7

Fiat Ansaldo Autoblinda AB 41 Italian Armoured Car Trials

In this report the armoured car in the attached photographs was wrongly referred to as an Autoblinda 40 armoured car: this has been corrected. The AB 40 was armed with twin 8mm Breda 38 machine guns in the turret and one in a ball mount in the rear of the hull. Later versions were armed with a single, more powerful, Cannone-Mitragliera Breda 20/65 Modello 1935

This Fiat Ansaldo Autoblinda AB 41 Italian Armoured Car was captured by the British in North Africa and reused against its previous owners prior to being sent for vehicle trials.

20mm gun and two 8mm Breda 38 machine guns. The first machine gun was coaxial to the cannon, on the left, in the turret and the second in a ball support on the rear of the vehicle. That version was given the designation Autoblinda 41.

The vehicle used in trials was one of a number of armoured cars captured by the British in North Africa and kept by the British Army for reuse against their former owners. The one in the report photographs was assigned to the Polish Carpathian Lancers Regiment. Polish mechanics worked on the armoured car and brought it back to a serviceable condition. It was issued to Jan Billewicz, Commander of the 1st Squadron and saw operational service with the regiment between May and August 1942. After the Polish Carpathian Lancers Regiment was released from the Nile Delta Barrage defences in Egypt this armoured car was shipped to England for tests. What follows is the report on those field trials, but there is confusion whether it was the AB 41 in the report photographs or a different vehicle, an AB 40 armed with three machine guns.

SECRET

FIGHTING VEHICLE PROVING ESTABLISHMENT, FIELD TRIALS REPORT ON ARMOURED CAR, AUTOBLINDA 40, ITALIAN 4X4 PERFORMANCE TRIALS

REFERENCE

FVPE Report No.FT.1595
DTD Project No.DE.10074
DTD File No.10/40/41

STATEMENT

It was required to ascertain the performance of a 4x4 Italian armoured car known as an Autoblinda 40, in order to provide information of a nature which might prove useful for comparison with other vehicles at present in use in the service. The vehicle had covered 7,280 kilometres (4,520 miles) according to the odometer prior to these trials taking place.

DESCRIPTION OF VEHICLE

A very full and concise description of the vehicle is contained in a preliminary report prepared by the school of tank technology and published in May 1943. Briefly, the vehicle is a petrol engined medium armoured car incorporating 4 wheels, all of which are steerable, independently sprung and driven from a central transfer box by propeller shafts direct to the wheel hub reduction gears. The six-cylinder engine of an estimated 100 to 120bhp, drives through a dry plate clutch to a 5 speed and overdrive gearbox and thence through a forward and reverse gearbox into the transfer box.

A steering wheel and a set of controls are provided in the rear driving position, but for this driver's seat only the first four gears can be engaged. The armament consists of one 2cm Breda and one 8mm Breda machine gun mounted in the turret, and one 8mm Breda machine gun mounted in the hull next to the rear driving position. The armour thickness is between 6mm to 9mm.

CONDITIONS

Estimated battle weight
Front Axle = 3 tons 7cwts
Rear Axle = 4 tons 3cwts
Total = 7 tons 10cwts

MILEAGE

Road	Cross Country	Alpine Course	Hungry Hill	Total
167	22	2	2	193

OBSERVATIONS

1. Repairs–Prior to any running, a number of minor repairs had to be carried out, including an overhaul to the steering layout. The front steering box sector was loose on its shaft, one of the securing bolts having sheared, and one nut being adrift. In the rear steering box, all four of the sector-securing bolts had sheared (it is noteworthy that in order to service the rear steering box, the engine had to be removed). All the steering linkage ball pins had excessive play, the front drop arm was loose on its shaft, and the steering arms were loose in the hubs. All necessary repairs were carried out to the steering and new primary and secondary cups were fitted to the brake master cylinders. The hand transmission brake lining, which was completely worn out, was not renewed. Even after these repairs had been carried out, the steering arms continually worked loose in the hubs and caused wheel wobble, which was more pronounced on the rear than on the front wheels.

2. Handling of Vehicle–Entry to the front driver's seat is very awkward due to the interference of the turret gunner's seat support tubes. Vision for the front driver is good forwards, but the lack of vision ports in the hull sides severely curtails lateral vision. The rear driver has a moderate field of view rearwards, but again suffers from a lack of lateral vision, unless the side entrance doors are open.

The seating position for the front driver is comfortable, with all the controls handy excepting the differential locking lever, which is situated under the floorboards beneath the turret and could only be operated by a member of the turret crew. The only instruments supplied in front are a fuel gauge and speedometer; oil pressure and water temperature gauges are fitted to the rear driver's control panel.

The clutch operated smoothly, and the brakes worked fairly well with no unduly pedal pressure. Gear changing of the crash type gearbox did not prove difficult, although the lack of a clutch stop precluded rapid upshifts.

3. Performance Trials—No difficulties were experienced during these performance trials, except that the vehicle showed symptoms of becoming uncontrollable during the flying quarter mile runs, due to a combination of wheel wobble and oversteering. When compared to a new armoured car, Daimler II, of similar battle weight and running under similar conditions, the performance figures show that the Autoblinda is:

Better than the Daimler II Armoured Car on	Worse than the Daimler II Armoured Car on
Stalling.	Overall length.
Turning circles.	Wheel base.
Cooling at max. torque.	Angle of overturn.
Cooling on high-speed road run.	Average speed on high-speed road run.
Cooling on cross country run.	Tilting (fuel leakages)
Fuel consumption on high-speed run.	Distortion.
Fuel consumption on cross country run.	Brakes generally.
Speed on cross country run.	Flying ¼ mile.
Fuel capacity.	Standing ¼ mile.
	Climbing a 24° concrete slope in reverse.

Exactly similar figures were obtained on Miles Hill gravel slopes and on the 24° concrete slope with vehicles proceeding uphill.

FUTURE ACTION

(1) The vehicle is being retained for pitch and roll trials under Project No. DE.10073.

(2) It is recommended that this project be closed.

SIGNED

Major J. W. Mills for OIC Field Trials
and Brigadier W. F. Morrogh Commandant, FVPE Chertsey
13 February 1945.

APPENDIX 'A'

AUTOBLINDA TYPE 40 (BUILT 1941) PERFORMANCE TRIALS

1. AXLE WEIGHTS

Front Axle = 3 tons 7 cwts
Rear Axle = 4 tons 3 cwts
Total = 7 tons 10 cwts

2. OVERALL MEASUREMENTS

Length	17ft 2 ½in	to front towing attachment
Width	6ft 4 ⁷⁄₈in	to mudwings
Height	7ft 11in	to top of aerial base
Track	5ft 6in	(taken to C/L of tyre path)
Wheelbase	10ft 6in	(taken to C/L of wheels)

3. BELLY CLEARANCE

Front = 13 inches
Rear = 12 ¹⁄₈ inches

4. CENTRE OF BALANCE

From C/L of front wheels   5ft 9 ½ inches
From C/L of rear wheels   4ft 8 inches

5. ANGLE OF OVERTURN

Left side up 38°
Right side up 38° 30'

6. TURNING CIRCLES

	To outside of wheels	Radius to C/L of vehicle
Right hand	40ft 6 ½ inch diameter	17ft 9 3/8 inch
Left hand	37ft 10 ½ inch diameter	16ft 5 3/8 inch

7. TYRE SIZES, TYPE AND PRESSURES

Three Pirelli Superflex 9.75 x 24 Libia made in Italy.
Three Pirelli Superflex 9.75 x 24 Importe D'Italia (E).
Front = 44lb/sq. in.
Rear = 52lb/sq. in.

8. COOLING TRIAL, MAXIMUM TORQUE

Speed = 6.9 mph
Gear used = 3rd

Maximum water T.D. into radiator	116°F	After 35 minutes
Maximum water T.D. out of radiator	99°F	After 30 minutes
Maximum engine oil T.D. into cooler	70°F	After 35 minutes
Maximum engine oil T.D. out of cooler	52°F	After 35 minutes
Maximum gearbox oil T.D	182°F	After 5 minutes
Maximum transfer box oil T.D	88°F	After 40 minutes

9. 50 MILES ROAD RUN

Traffic conditions–fairly heavy
Conditions of road–dry

Maximum water T.D. into radiator	97°F	After 50 minutes
Maximum water T.D. out of radiator	86°F	After 50 minutes
Maximum engine oil T.D. into cooler	97°F	After 60 minutes
Maximum engine oil T.D. out of cooler	60°F	After 90 minutes
Maximum gearbox oil T.D	183°F	After 80 minutes
Maximum transfer box oil T.D	111°F	After 100 minutes

Average speed   28.6mph
Average fuel consumption   5.88mpg
Average engine oil consumption   50mpg

10. 22 MILES CROSS COUNTRY

Gears used–3rd, 4th and 5th
Conditions of course–Very hard and frozen ruts

Maximum water T.D. into radiator	78°F	After 10 minutes
Maximum water T.D. out of radiator	68°F	After 10 minutes
Maximum engine oil T.D. into cooler	49°F	After 30 minutes
Maximum engine oil T.D. out of cooler	47°F	After 50 minutes
Maximum gearbox oil T.D	181°F	After 10 minutes
Maximum transfer box oil T.D	77°F	After 50 minutes

Average speed   16.9mph
Average fuel consumption   4.4mpg
Average engine oil consumption   44mpg

N.B. Thermostats removed

11. CIRCUIT OF ACTION

Main fuel tank over driver's wheel   capacity 12 gallons 4 pints
Main fuel tank under driver's seat   capacity 27 gallons 4 pints
Auxiliary fuel tank   capacity 4 gallons 4 pints
Total   capacity 44 gallons 4 pints

Road = 262 miles
Cross country = 196 miles

12. MILES HILL SLOPES

(a) Concrete 'A' (1:2.25 = 24°)
     Condition of surface   Dry.
     Straight climb, Forward   Successful.
     Straight climb, Reverse   Failed. Lack of torque.
     Stop and Restart, Forward   Successful.
     Stop and Restart, Reverse   Failed. Lack of torque.
(b) No.4 gravel slope (1:2.43 = 22.33°)
     Condition of surface   Thinly frozen deep mud and rutted.
     Straight climb, Forward   Failed. Lack of torque.
     Straight climb, Reverse   Failed. Lack of torque.
     Stop and Restart, Forward   Failed. Lack of torque.
     Stop and Restart, Reverse   Failed. Lack of torque.
(c) No.5 gravel slope (1:2.74 = 20°)
     Condition of surface   Frozen and slightly rutted.
     Straight climb, Forward   Successful.
     Straight climb, Reverse   Successful.
     Stop and Restart, Forward   Successful.
     Stop and Restart, Reverse   Successful.

13. TILTING

(a) Vehicle facing up 24° slope.
     Oil pressure (idling)   20lb/sq.in
     Oil pressure (maximum rpm)   34lb/sq.in
     Leakages   Petrol leaking from main tank filler cap.

  (b) Vehicle facing down 24° slope.

Oil pressure (idling)	20lb/sq.in
Oil pressure (maximum rpm)	39lb/sq.in (quantities of blue smoke)
Leakages	Nil.

  (c) Vehicle across 15° slope. Left side up

Oil pressure (idling)	16lb/sq.in
Oil pressure (maximum rpm)	39lb/sq.in (burnt oil smoke from exhaust)
Leakages	Nil.

  (d) Vehicle across 15° slope. Right side up

Oil pressure (idling)	16lb/sq.in
Oil pressure (maximum rpm)	41lb/sq.in (blue smoke from exhaust)
Leakages	Fuel leaking from main fuel tank filler cap.

  (e) Vehicle on level ground

Oil pressure (idling)	34lb/sq.in
Oil pressure (maximum rpm)	41lb/sq.in
Leakages	Nil.

Note: Above pressures recorded with a cylinder head water temperature of 55°C.

14. DISTORTION (WHEEL RAISED UNTIL ANOTHER JUST LEAVES GROUND)

(a) Left front wheel	(i) No lock.	Offside rear wheel fouls mudwing.
	(ii) Left lock.	Successful.
	(iii) Right lock.	Offside rear wheel fouls mudwing.
(b) Left rear wheel	(i) No lock.	Nearside front wheel fouls mudwing.
	(ii) Left lock.	Offside rear wheel fouls mudwing.
	(iii) Right lock.	Nearside front wheel binding hard on mudwing.
(c) Right front wheel	(i) No lock.	Nearside front wheel fouls mudwing.
	(ii) Left lock.	Offside front wheel fouls mudwing. Nearside rear fouling hard.
	(iii) Right lock.	Nearside rear wheel fouls mudwing.
(d) Right rear wheel	(i) No lock.	Successful.
	(ii) Left lock.	Successful.
	(iii) Right lock.	Nearside front wheel fouls mudwing.

15. BRAKES

  (a) Retardation on dry level concrete

     (i) Hand–mean of 4 runs at 20 mph = Transmission brake not used as it was worn out.

    (ii) Foot–mean of 4 runs at 30 mph = 16.2 f.s.s.

(b) Controlled descent of Beacon Hill (Ave. gradient 1:10,515 yards), gears in neutral.
  (i) Hand–holding vehicle to 8 mph = Not used as transmission brake was worn out.
  (ii) Foot–holding vehicle to 8 mph = Successful. All drums warm.
(c) Controlled on concrete gradient 1:2.25 (24°) gears in neutral.
  Hand–facing uphill = Not used as transmission brake was worn out.
  Hand–facing downhill = Not used as transmission brake was worn out.
  Foot–facing uphill = Held.
  Foot–facing downhill = Held.

16. STALLING

Gear used = 3rd
Speed = 6.9mph
Gross retardation to stall engine = 12.7 f.s.s.

17. ASCENT OF BEACON HILL (AVERAGE GRADIENT 1:10,515 YARDS)

Standing start: speed 13.9mph, gears used 2nd and 3rd
Flying start: speed 15.5mph, gears used 5th, 4th and 3rd

18. SPEEDS

(a) ¼ mile standing start (mean of 2 runs) 21.35 mph
(b) ¼ mile flying start (mean of 2 runs) 44.6 mph

19. TOWING HOOKS

(a) When towing a vehicle of similar weight across country. Successful.
(b) When being towed across country. Successful.

This captured Fiat Ansaldo Autoblinda AB 41 Italian Armoured Car was assigned to the Polish Carpathian Lancers Regiment. Polish mechanics worked on the armoured car and brought it back to a serviceable condition. It was issued to Jan Billewicz, Commander of the 1st Squadron and saw operational service with the regiment between May and August 1942.

The Fiat Ansaldo Autoblinda AB 41 Italian Armoured Car was armed with a Cannone-Mitragliera Breda 20/65 Modello 1935 20mm gun and two 8mm Breda 38 machine guns. The first machine gun was coaxial to the cannon, on the left, in the turret and the second in a ball support on the rear of the vehicle.

The Fiat Ansaldo Autoblinda AB 41 Italian Armoured Car was powered by a SPA 16 120hp petrol engine and it had an average road speed of 28.6mph.

The Fiat Ansaldo Autoblinda AB 41 Italian Armoured Car had a foldable radio aerial on the right side of the hull.

The Wilton-Fijenoord Pantserwagen Armoured Car

Why should you be interested in a Dutch armoured car built in 1934 that did not enter mass production? In fact, only three were built. The answer is that it was based on a Krupp chassis and engine and one saw service with the German Army in the Second World War during the Battle for Berlin.

The Dutch-built Wilton-Fijenoord armoured car was designed by the German firm Krupp.

German companies, like Krupp, were not allowed to build armoured vehicles for the German Army under the terms of the 1919 Versailles Treaty. In 1934, Adolf Hitler became Germany's head of state with the title of *Führer und Reichskanzler* (Leader and Chancellor of the Reich). In speeches he blamed a lot of the ills suffered by the German people after the First World War on the Versailles Treaty.

The Nazi regime started to publicly ignore the terms of the Treaty and rearm. German companies and tank designers had been working in secret since the 1920's in places like Sweden and Holland on new AFV designs. Senior German military officers visited other European armies on official visits to study new concepts in mechanised warfare. They were clandestinely preparing for war. The development history of the Dutch Wilton-Fijenoord Pantserwagen armoured car is a good example of the pre-World War Two German involvement in military weapons construction. They learnt design lessons from this experience that helped them build more effective armoured cars for the Wehrmacht.

During the First World War the Dutch remained neutral, but they were affected by the War. Its strength of its armed forces had gradually dwindled since the Napoleonic wars. Conscription had to be introduced, and troops were mobilised to protect the country's ports and borders. Food shortages became common as the country was surrounded by warring states.

Dutch Army senior officers drew lessons from the Great War. They realised that their army needed to transition to a mechanised, more mobile force. The off-road capabilities of bullet proof, armoured vehicles that could engage enemy infantry and strong points was noted. The problem was the lack of money.

The Great Depression following the 1929 Wall Street Crash led to widespread poverty and unemployment in Holland. The Dutch defence budget was only increased in 1936 after Germany remilitarised the Rhineland. When Hitler annexed Austria and occupied the Czech Sudetenland, the budget was further increased in 1938. It was too little, too late. Although Dutch official policy was still neutrality, the Dutch Government desperately tried to buy new arms, equipment, ammunition and tanks. A lot of the ordered weapons never arrived in time. On May 10, 1940, the Germans invaded Holland, Belgium and France. The Dutch Army fought hard with the resources and weapons they had. In many locations the German Army could make little or no headway. Following the bombing of Rotterdam, where 900 civilians were killed and 78,000 were left homeless, plus the threat to do the same to the city of Utrecht, the Dutch capitulated on 15 May 1940.

Before the Wall Street Crash, the Royal Dutch Army (*Koninklijke Landmacht*) had developed a tactic that used machine gun armed armoured cars to give the infantry close support in attack and defence. They started looking at obtaining new armoured vehicles. The Wall Street Crash imposed financial restraints on the Dutch Government, and this reduced the military's ability to obtain new fighting machines, but this was not the situation with the Royal Netherlands East Indies Army (K.N.I.L.).

In 1602 the Dutch East Indies company built their first fortress in Jakarta on the island of Java. This was the start of Dutch colonial expansion in the area. The Dutch East Indies territory constantly continued to increase in size to what is now modern-day Indonesia. In the 1930s the K.N.I.L. required armoured vehicles for a military policing role in case of insurrection and guerrilla warfare by Indonesian Nationalists. There was also concern about Japanese expansionism.

On 16 August 1933, the Dutch Ministry of Colonies (Ministerie van Koloniën) ordered three armoured cars, priced at 25,500 Dutch Guilders each, to be constructed for the K.N.I.L. by the Dock and Shipyard company Wilton-Fijenoord Limited, based in Schiedam. A further order was promised to be made in the following year for three more armoured cars, following the successful delivery of the first batch and successful trials. Wilton-Fijenoord had expertise in shipbuilding and repair, not in designing and building armoured vehicles. However, as they had built armoured ships for the Navy, they did have experience working with armour plate.

The German firm Krupp, from Essen, designed the vehicle and Wilton-Fijenoord built it. This was an ideal opportunity for Krupp, as they could learn from the experience of seeing their design being built and tested by the Dutch. These armoured cars were based on a six-wheeled 6x4 Krupp-Protze L2H43 truck chassis. Three chassis arrived in Schiedam on 18 November 1933, and construction commenced. The vehicle was powered by a Krupp M304 4-stroke, 4-cylinder, air-cooled 60hp petrol engine that was horizontally opposed. Cooling was an issue as it was going to be used in the tropics. The engine was fitted with cooling rings along which around 1,000 litres of air per second was blown by a compressor. It had four forward and one reverse gear. Engine power was transmitted by a single disc clutch and an Aphon gearbox. For better cross-country performance it was equipped with an additional gear change. It had a maximum road speed of 43.5mph (70km/h). The 60-litre capacity fuel tank gave the vehicle a maximum operational road range of 186 miles (300km).

Armoured cars are used for a variety of tasks. In the reconnaissance role they would advance until contact and then retreat out of harm's way so the crew could provide intelligence of the location and size of enemy forces.

By the standards of the early 1930s, the Wilton-Fijenoord Pantserwagen Armoured Car was fitted with four pieces of advanced technology. The first piece of advanced technology was the fact that this vehicle had two driving positions. The driver sat on the left side of the vehicle at the front as normal for a German and Dutch truck, but a second set of driver's controls were fitted at the rear of the vehicle. This meant that in an emergency one of the crew could take over the controls and drive the vehicle safely backwards whilst sitting at the back of the vehicle. The ability for the vehicle to drive back the way it came without having to waste time turning around is a great advantage. It is particularly useful in urban environments on narrow roads. This was an important feature, as its turning circle was poor as it required 4.4 metres. Unusually the driver's controls at the rear of the vehicle were on the right side and the rear machine gun was next to him on the left side. I was expecting the rear driver's controls to be on the left side of the vehicle as would be standard practice on European manufactured vehicles. I have not found out the reason why this was designed in this fashion.

The second piece of advanced technology was kept in the two boxes above the back wheels, on both sides of the vehicle. This is where two track loops were stored. When it was necessary to drive cross-country the crew could loop the track over the two rear wheels transforming the Wilton-Fijenoord Pantserwagen Armoured Car into a half track, resulting in better off-road performance. The four back wheels were equipped with hydraulic brakes. The tires were made of solid rubber, which made them bulletproof.

Around the same time, the British Mechanisation Experimental Establishment (MEE) in Farnborough was evaluating a variety of six-wheeled lorries that could be fitted with looped tracks over the two rear wheels on their cross-country course. The problem with this design was that the track had a tendency to fall off the wheels in very undulating terrain. The idea had spread to most European military forces, but by the beginning of the Second World War the concept was superseded by the design of purpose-built half-track armoured vehicles that had track guide lugs, holes in the track that fitted in a sprocket wheel, and were fitted with a way of tightening the track tension.

Modern critics look at this armoured car design and dismiss it because it was only armed with machine guns and did not have the ability to fire armour piercing shells, but it was not intended to engage in combat with tanks. It had to protect itself from enemy infantry and armed Indonesian Muslim Nationalist insurgents. The Wilton-Fijenoord Pantserwagen Armoured Car was armed with three machine guns of which two were mounted in the hull and one in a fully rotating turret. The exact type of machine gun has not been found specified in the existing documentation. Surviving operational images show the vehicle was fitted with armoured covers for the machine gun barrels but most of the archive photographs are of vehicles that have not been fitted with machine guns. The protective armoured covers do not have a machine gun barrel poking out the front.

I smiled when I first saw the photograph of the fighting compartment. I am used to seeing images of powered or manual hand wheel turret traverse systems. This was the first time I had seen a turret traverse system powered by pedal power. Well, what should I have expected. This is a Dutch armoured car, and the Dutch love their bicycles. It is only natural that their army's future armoured car turret traverse was turned by pedalling. This idea was not as silly as it initially looked. It enabled the commander/gunner to use both hands to operate the turret machine gun.

The machine guns in the hull had a traverse angle of 25° to each side. A half-ball shaped gun mantlet provided bullet splatter protection for the gap where the gun shield met the opening in the hull armour. The crew has been shown in some sources as consisting of three men. This could be problematic in a combat situation. The commander/turret gunner would always have to stay in the turret. The forward driver could not operate the forward hull machine gun whilst he was driving. A third man would have to operate the machine gun. When the vehicle had to retreat in a hurry a fourth man was required to operate the rear driving controls, but he could not fire the rear gun whilst doing so. That required a fifth man. To operate the Wilton-Fijenoord Pantserwagen Armoured Car effectively in a combat zone it needed a five-man crew.

The third piece of advanced technology used on this armoured car design is that the exterior of the vehicle could be electrified so as to deter enemy infantry or rioters from trying to climb on to the armoured car. Another self-defence weapon, to help the crew retreat in safety, was the fitting of a small hatch installed in the bottom plate. The crew could use this hatch to throw out onto the floor crowd control Lachrymatory tear gas grenades. If the armoured car was going to be used in a live fire combat zone where there was a possibility that it would come under air-attack an additional anti-aircraft machine gun could be mounted on top of the turret.

The fourth piece of 1930s advanced technology was the use of sloped armour. To reduce the weight of the vehicle, where possible, the designers placed the armour plates at an angle. Apart from the armour plate covering the six wheels most of the vehicle surfaces were angled. Thinner armour could be used, if set at an angle, to give the same protective qualities as thicker, heavier, vertical armour. The armour was riveted together rather than being welded.

During firing trials, the armour was found to be resistant to 7.9 mm S.M.K. (Stahlmantelkern) bullets from a range of 30 metres. Another advantage of having sloped armour was that it reduced the chance that a thrown grenade could land and stay on top of the vehicle, causing damage.

There were large crew hatches in the vehicle hull, one on each side. If one hatch was damaged or obstructed the crew could escape their vehicle by using the other side hatch. The commander/gunner had a hatch in the roof of the turret. The driver's vision port could be locked in the open position, but in combat areas it was closed and locked. He could see through a vision slit in the hatch protected by bulletproof glass. He had to rely on the commander and other crew members to give him driving instructions as his vision left and right was very poor. The driver's position at the rear of the vehicle had a similar vision port. The commander had a forward-facing periscope, and five vision ports around the turret walls. This gave him a good view of his surroundings. Central headlights were installed on the front and the back of the vehicle. The one at the front could be folded into the superstructure. The one at the rear was protected by an armoured flap which could be raised when the armoured car was being driven backwards. Eight lamps were installed inside the fighting compartment.

By the beginning of April 1934, the first Wilton-Fijenoord Pantserwagen Armoured Car was ready to be assessed. On 12 April 1934, the vehicle crew, and a Dutch military detachment, including two KNIL Captains, met with representatives from Krupp in the Dutch city of Roosendaal. The vehicle successfully passed the trials with no major problems. On 26 April 1934, it was shipped to the Dutch East Indies, arriving in the port of Tanjung Priok at the end of May 1934. The second vehicle arrived in the Indies on 21 June 1934.

The KNIL ran their own vehicle trials. In October 1934 they officially rejected the vehicle as not fit for service with the KNIL. The Krupp engine could not run on standard petrol. It required higher octane petrol that was not available in the Dutch East Indies. They also identified a structural weakness that caused the spring between the chassis and the wheels to break. They found that the power to weight ratio was not enough; it had poor acceleration and climbing capabilities. Different terrain, climate and fuel quality meant that the armoured car did not perform as well as it did in Holland. The delivery of the third armoured car was postponed and further orders cancelled.

In February 1935 the two vehicles in the Dutch East Indies were sold to Brazil where they were used by the Police Force of São Paulo. In May 1940, the third vehicle was captured by German troops and assigned to the internal public order and security *Ordnungspolizei*. The Germans used it eventually in the defence of the Reichs Chancellery internal patio during the Battle of Berlin in 1945, where it was destroyed by Soviet forces.

Above: This is the rear view of the Wilton-Fijenoord armoured car at the 'Korps Rijdende Artillerie'. Note the big, large stowage box above the two rear wheels that were used to store the track loops that fitted over the rear two wheels.

Opposite above: This is the front view of the Wilton-Fijenoord armoured car at the 'Korps Rijdende Artillerie'. Note the big, retractable headlight. The driver's position was on the left side of the vehicle next to the machine gun.

Opposite below: Without the wheels fitted the amount of sloped armour surfaces can be more easily be appreciated.

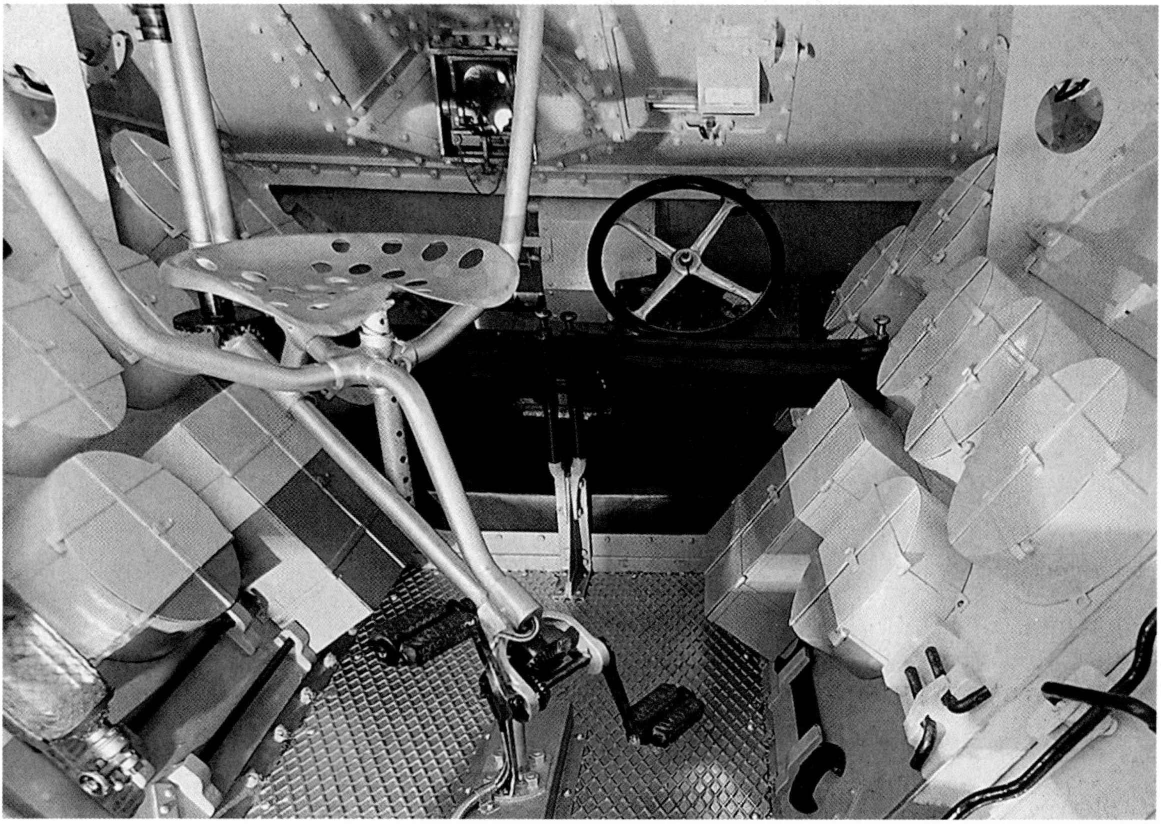

Opposite above: This photograph taken at the Wilton-Fijenoord workshops shows the three vehicles in different stages of construction.

Opposite below: Rear driver's controls and machine gun can be seen. The bicycle pedals used to turn the turret are visible at the bottom of the photograph. Cogs and wheels are used rather than a bicycle chain.

Below: View of the inside of the turret including the commander's seat and vision ports.

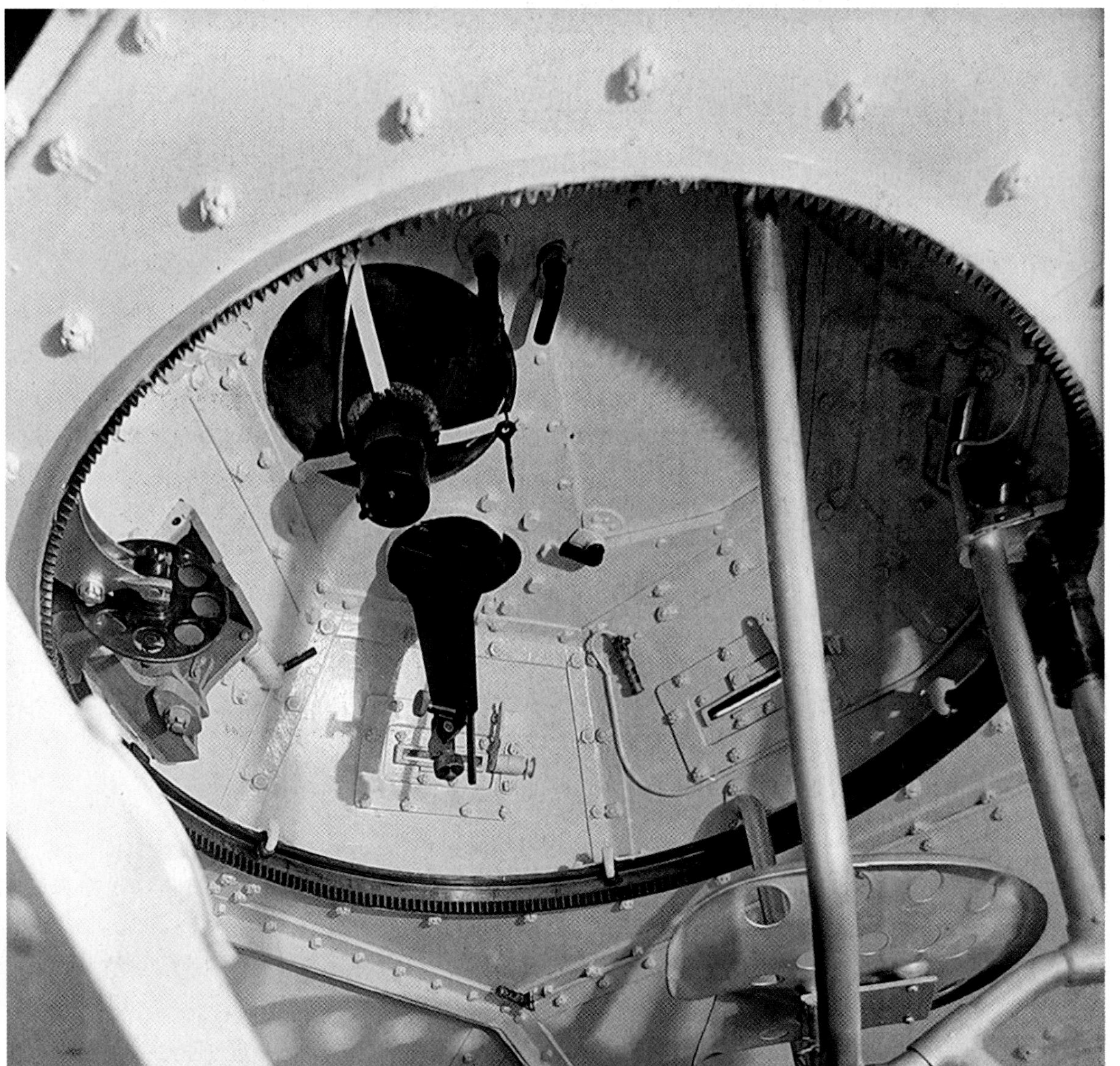

Above: The only vertical pieces of armour on the Wilton-Fijenoord armoured car were used to protect the wheels.

Left: Three hatches were built into the front armour to make maintenance easier.

The front driver's position.

This is a photograph of a Wilton-Fijenoord armoured car, together with the improvised Ehrhardt armoured car in Arnhem, captured by the German Army in May 1940.

One of the Wilton-Fijenoord armoured cars photographed during testing in the Dutch East Indies.

Two Wilton-Fijenoord armoured cars in Brazil during a police parade, April 1936.

This photograph shows a destroyed Wilton-Fijenoord armoured car at the Reichs Chancellery in Berlin. In front of it is a destroyed Schupo-sonderwagen Daimler/21 DZVR armoured police car. Notice the stowage box is open over the rear two wheels and contains tracks. They were to go over the rear two wheels to turn the vehicle into a half-track when it needed to drive over rough terrain cross country.

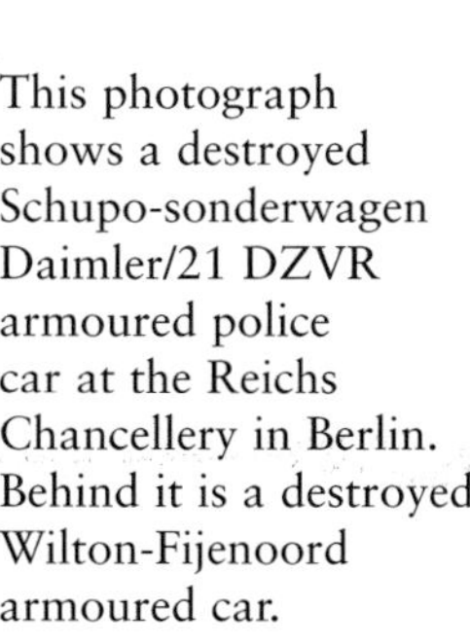

This photograph shows a destroyed Schupo-sonderwagen Daimler/21 DZVR armoured police car at the Reichs Chancellery in Berlin. Behind it is a destroyed Wilton-Fijenoord armoured car.

9

The Mouse: A German One-Man Light Tank Proposal

A bundle of plans and documents went on sale at an auction house. They appeared to show previously unseen 'blueprints' for a German one-man light tank design that had been submitted to Albert Speer, the German Minister for Armaments and War Production, by a German officer. As no supporting documents that mentioned this tank had been found in the archives, these documents were viewed as possibly fake.

In November 2019 C & T Auctioneers and Valuers Ltd, based in Ashford, Kent, uploaded photographs on their website of plans, sketches and official documents relating to the same one-man light tank that were to be sold on 11 December 2019 at their auction room. They put a value on them at £200 to £500. These were not the same documents that had been offered for sale three years previously.

Military historians from Germany, America, Holland and Britain started to take more interest in this tank design. It was discovered that the style of

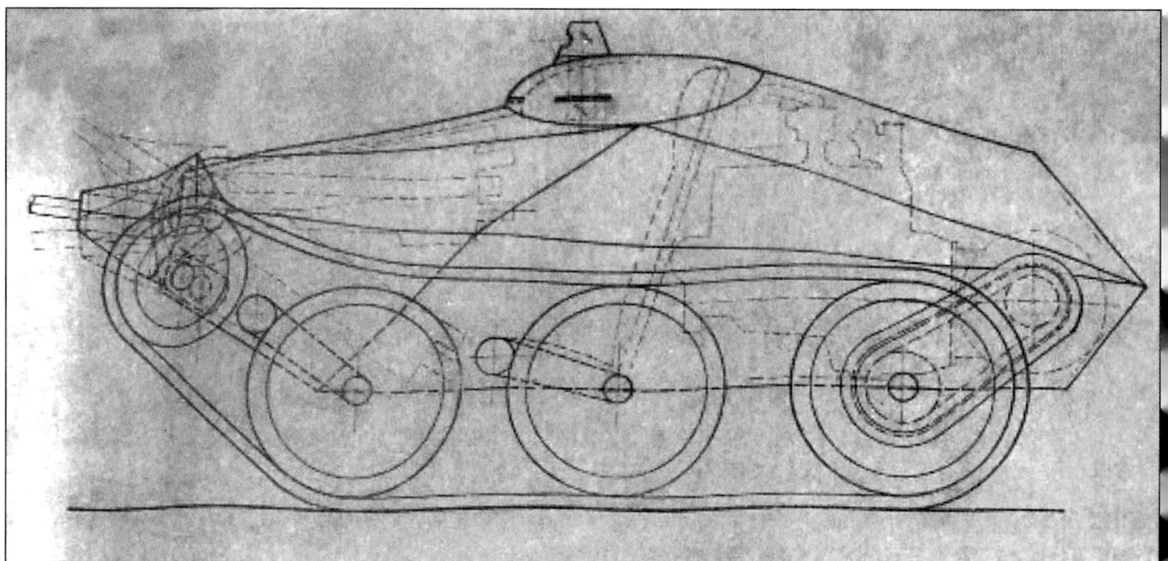

Side view of Leutant Franz-Georg Gmelin's one-man tank design.

words used on the tank plans were written in an old German dialect called *Sütterlinschrift*. The *Sütterlin* handwritten scripts were introduced in Prussia in 1915. The Nazi Party banned *Sütterlin* typefaces in 1941 as they were seen as chaotic and replaced them with Latin-type letters like Antiqua. However, many German speakers brought up with this writing system continued to use it well into the post-War period. *Sütterlin* was taught in some German schools until the 1970s, but no longer as the primary script.

The tank design was submitted by Leutnant Franz-Georg Gmelin. The officer's name was written in the old way with the surname first. The address he gave was full of abbreviations: Issing b. Landsberg a.L. (Issing near the town of Landsberg at the River Lech). Issing is now part of the town of Landsberg, which is 37 miles (59km) west of Munich. It is not known why he was staying at home in Issing. He may have been injured and recovering from wounds. In documentation this address was referred to as his wife's address and not his 'field' address.

On the plans, the tank was called *'Entwurf zum 1 Mann–KleinpanzerKampfwagen'*, which translates to: 'Designed for one man–small, armoured combat vehicle'. A better translation would be a 'one-man light tank' or 'one-man tankette'. On the typed documents the tank was called *'1–Mann–KleinpanzerKampfwagen–Maus'*. The word *'Maus'* is German for 'mouse'. This could cause confusion as the German army already had a super heavy tank design ironically called the *'Maus'*.

The tank's dimensions were going to be 3.0m (9ft 10in) long, 1.9m (6ft 2.8in) wide and 1.3m (4ft 1.8in) high. The tank's armour would range from 10mm–38mm (0.39–1.49in), angled at 14-45°. The total combat ready weight was designed to be 2.7 tonnes with a ground pressure reading of 0.4kg/cm². An exact engine is not specified. The document states that a commercially available truck engine should be used. It was hoped that the top road speed would be 60km/h (37.28mph).

On the matter of the tank's armament the information in the documentation was not specific. It stated that it could be armed with fully automatic weapons up to 2cm and semi-automatic weapons up to 3.7cm. It went on to state that future planned versions could be armed with grenades, a flamethrower, smoke dischargers, a Panzer-Büchse (heavy anti-tank rifle), a Panzerschreck (hand-held shoulder launched rocket launcher used as an anti-tank weapon) and a Panzerfaust (hand-held single shot, recoilless anti-tank weapon).

The tank did not have a turret, but the crewman appears to have a cupola with three vision slits, out of which he had to see where he was going, look out for threats and find targets. The domed cupola was not circular and did not move. It appears to be made of cast metal. The rest of the armour would be cut from flat rolled armour plate. An armoured cover appears to project out of the cupola to protect the gun sight periscope. Two guns were to be mounted next to each other in ball mounts and projected out of the nose of the tank. It appears from the drawings that the gun on the right has a longer barrel suggesting that it was a more powerful weapon. The documentation states the guns would have a limited traverse arc of 30°.

The crewman sat in the centre of the tank. The engine was at the rear of the tank behind the crewman's seat. It powered the final drive and track drive sprocket wheel at the rear of the tank. This is an unusual configuration as most Second World War German tanks had the track drive sprocket wheels at the front of the tank. The track idler wheel was at the front and there were two large road wheels behind it. They were the same diameter as the track drive sprocket wheel. The tank was not fitted with track return rollers.

The tank tracks do not project in front of the two-gun barrels. In most circumstances this is not a problem as obstacles like a bank of earth, a rock or tree trunk could be driven over. The problem would come when the tank is driven down the side of a steep embankment, trench, or shell crater. There would be a danger that the front of the gun barrels would dig into the earth and be damaged.

Leutnant Franz-Georg Gmelin received a proposal rejection letter, dated 13 November 1944, from the Ministry for Armaments and War Production at his home address in Bavaria. They thanked him for his detailed proposal and mentioned that they had returned his plans and documents.

One of the main reasons for the rejection of his idea was that the Schützenpanzerwagen–Sd.Kfz 251 armoured infantry halftrack had proven itself for years accompanying Panzergrenadiers into battle, so there is no tactical need to move from that vehicle to a one-man light tank.

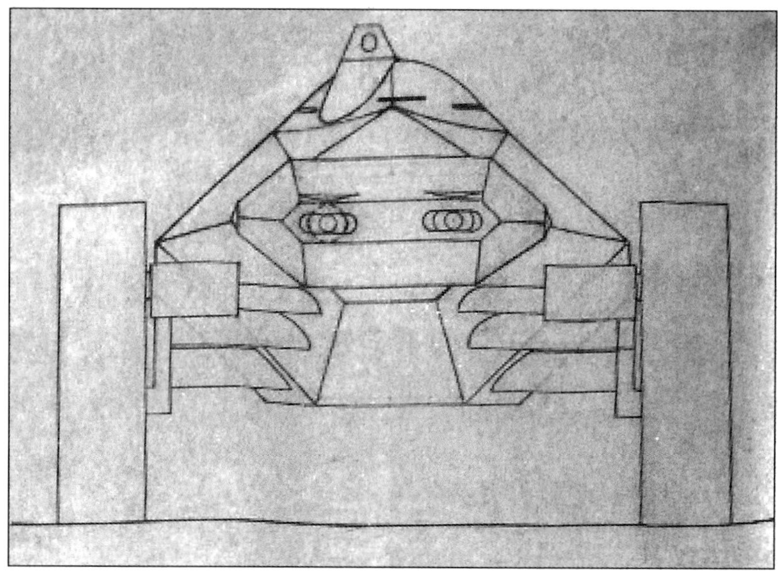

Front view of Leutant Franz-Georg Gmelin's one-man tank design.

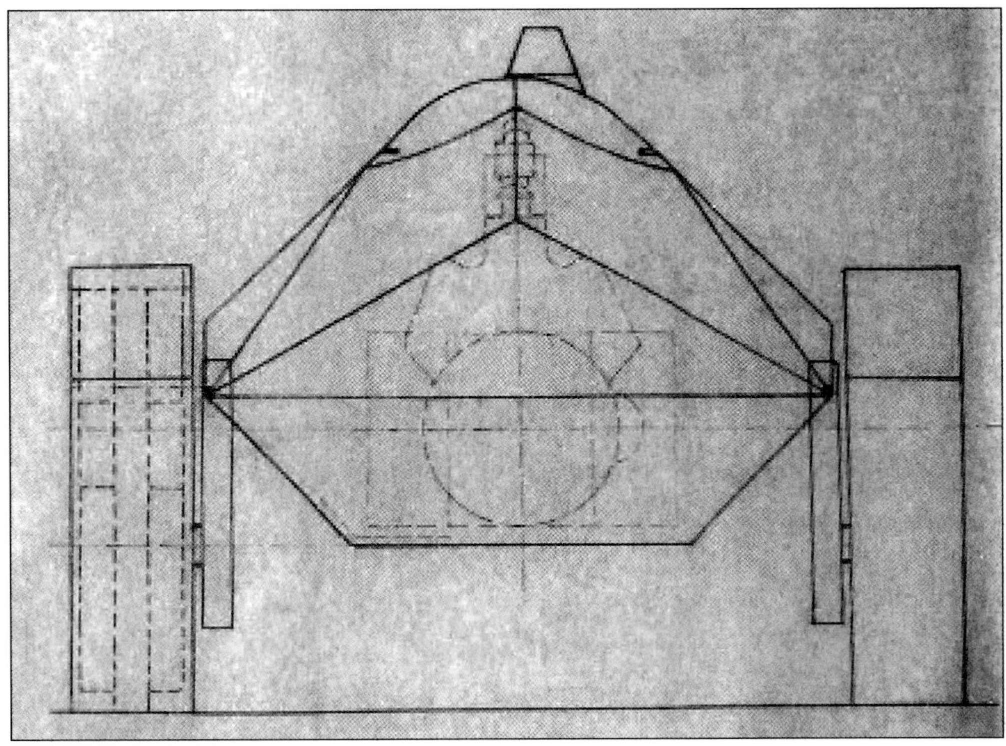

Rear view of Leutant Franz-Georg Gmelin's one-man tank design.

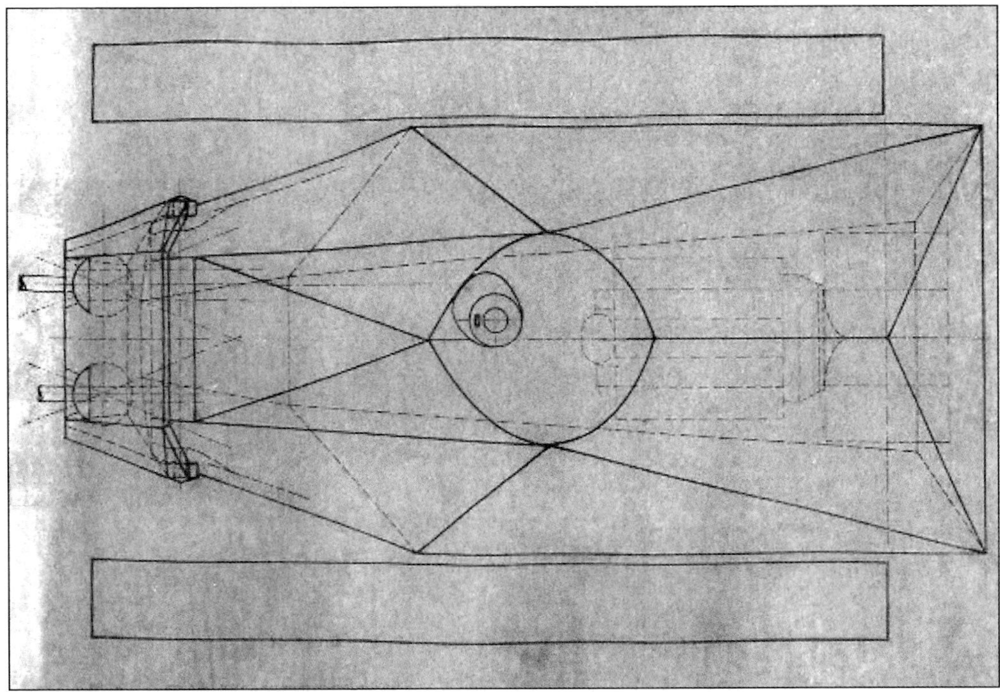

Top view of Leutant Franz-Georg Gmelin's one-man tank design.

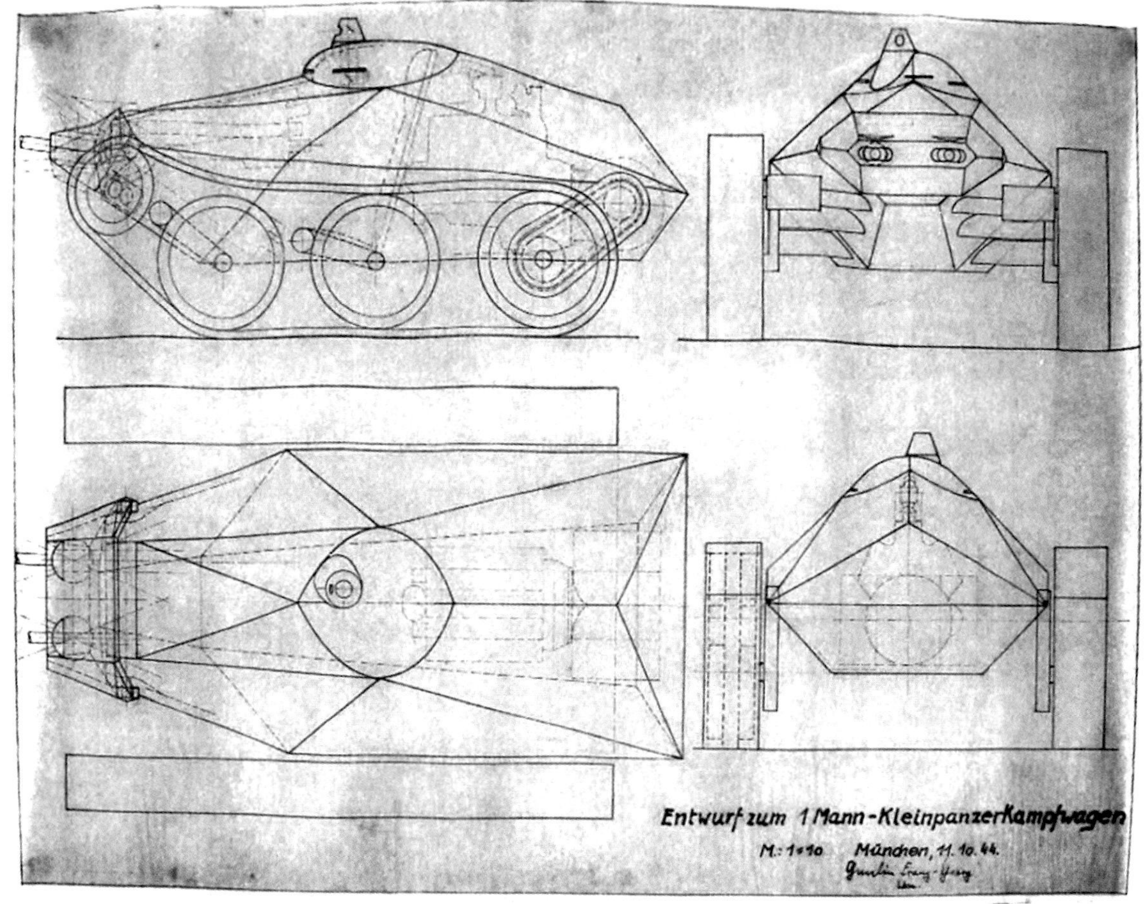

Full plan view of Leutant Franz-Georg Gmelin's one-man tank design.

10

Schildkröte Panzerspähwagen Light Amphibious Turtle Armoured Car

British Military Intelligence Section 10 (M.I.10) technical officers were part of T-Force. These units travelled just behind the front line. Their task was to grab German documents of a scientific technical nature, examine enemy vehicles, weapons and factories, plus capture scientists, weapons designers or senior armament construction workers before the Soviets. They found documentary evidence of the development of the Schildkröte Panzerspähwagen, a light amphibious armoured car that had been called the 'Turtle.' Two reports on this vehicle have been found in the National Archives at Kew.

German Light Amphibious Armoured Car Schildkröte Turtle I. Trippelwerke—Molsheim. This German drawing is dated 11 March 1942.

WAR OFFICE TECHNICAL INTELLIGENCE SUMMARY NO.179

13 JUNE 1945

AMPHIBIOUS ARMOURED CAR SCHILDKRÖTE (TURTLE)

GENERAL

Design work on this vehicle appears to have been started in 1942 as a private venture by Trippelwerke of Molsheim, a firm long associated with the development of amphibious cars. They appear to have experienced a great deal of difficulty in obtaining the necessary priorities to start production, and, having obtained the priorities, the components to which they applied. However, it appears that at least six prototypes were finally constructed, each differing from the others in some way or another. These were either unsuccessful in their trials or were not required by the Army, as quantity production was never started and the work on the 'Turtle' appears to have stopped some time ago.

ARMOUR

The armour appears to be flat rolled plate ranging in thickness from 5.5mm to 14.5mm with the exception of two bent plates forming part of each rear wheel casing. Armour thicknesses, layout and approximate angles to vertical are shown in the attached hull armour arrangement diagram.

ARMAMENT

Two versions of the 'Turtle' appear to have been tried, one armed with twin turret-mounted 7.92mm M.G.81 machine guns and the other with one 20mm M.G.151 main gun and a 7.92mm M.G.34 or 42 machine gun, also mounted in a turret with 360° traverse. The guns are electrically fired, and hand traverse and elevation of the turret and mounting is provided. The 20mm M.G.151 mounting incorporates a ring spring buffer. Ammunition for the 20mm M.G.151 is stowed in steel boxes measuring 151mm x 400mm x 650mm, each box containing 250 rounds.

ENGINE AND AUXILIARIES

The design called for an engine with a capacity of 2.75–3 litres, and there is evidence of two engines conforming approximately to this specification having been tried. The first engine was a 3.6 litre Opel, and the second a Tatra Model 87. The vehicle had a 24-volt electrical system. The gearbox was required to have three or four forward speeds, and that from the S.G.6 Model 41 was actually used. (The Trippel S.G.6 was a *Schwimmwagen* amphibious vehicle developed in the 1930s and used by the German ground forces during the Second World War.)

LAYOUT

The vehicle appears to be a conventional armoured, wheeled amphibian as regards layout. The turret is mounted centrally on the hull roof plate. The driver sits forward of the fighting compartment, in the centre of the hull. The engine is presumably mounted in the rear, behind the fighting compartment. The number of wheels is not known, although there is good evidence to believe that there are four, each being fitted with bullet-proof tyres. The means of propulsion in the water is not stated in the documents.

This drawing of the German Light Amphibious Armoured Car Schildkröte Turtle II was completed by British Military Intelligence Section 10 (MI10) Technical Officers, from information they discovered just after the war. It was dated August 1945. MI10 was responsible for weapons and technical analysis.

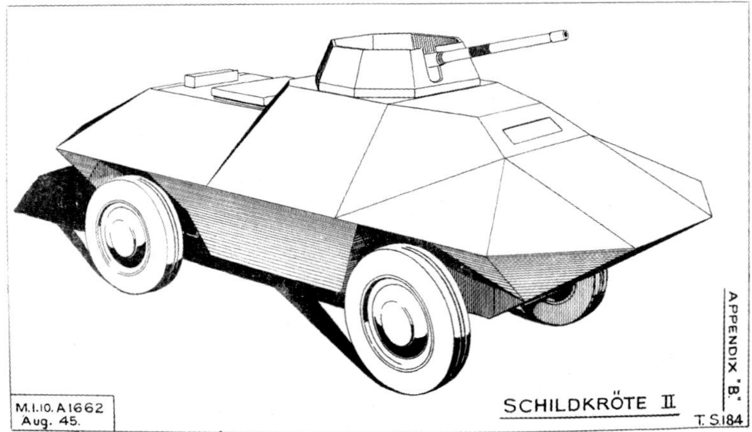

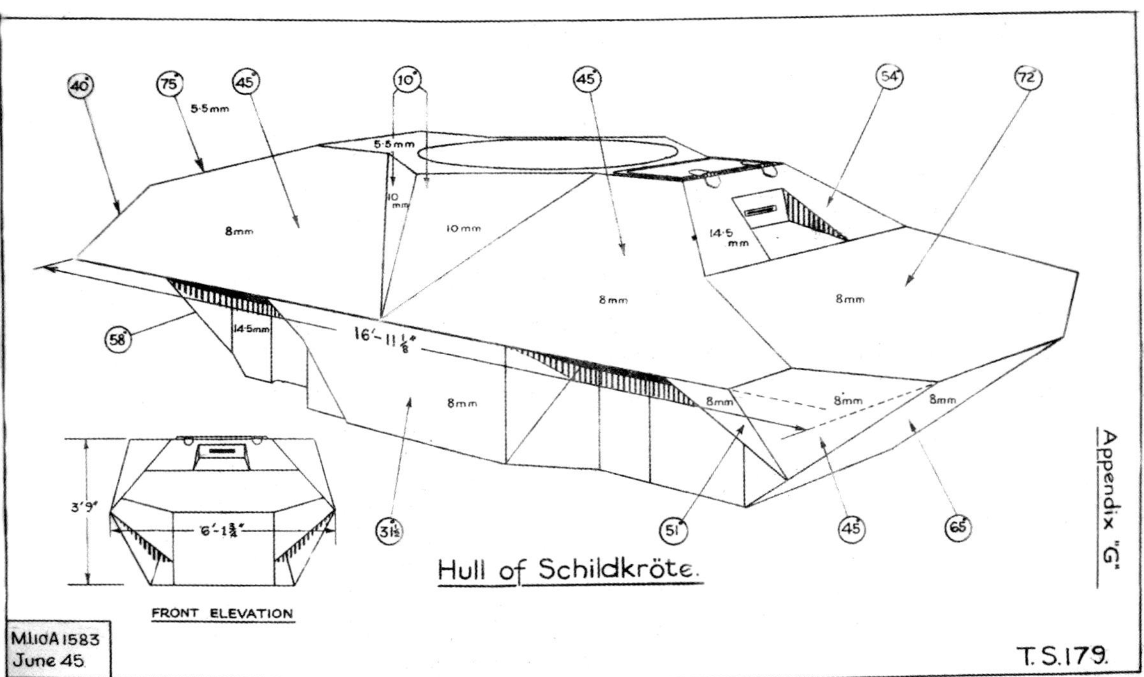

This drawing of the Hull of the Light Amphibious Armoured Car Schildkröte Turtle was completed by British Military Intelligence Section 10 (MI10) Technical Officers. It is dated June 1945.

WAR OFFICE TECHNICAL INTELLIGENCE SUMMARY NO.184

23 AUGUST 1945

LIGHT AMPHIBIOUS ARMOURED CAR SCHILDKRÖTE (TURTLE)

DEVELOPMENT

The 'Schildkröte' appears to be essentially an armoured and armed version of the S.G.6 amphibious car. The idea for such a vehicle appears to have been originated by Hans Trippel towards the end of 1941 and an order for one specimen was placed in January 1942. This vehicle was known as the Schildkröte I. Its characteristics were as follows:

S.G.6 chassis (the Trippel S.G.6 was a *Schwimmwagen* amphibious vehicle developed in the 1930s and used by the German ground forces during the Second World War.) Standard 6-cylinder Opel 2.5 litre 54hp engine, standard front and rear suspension steering system and 1mm to 1.5mm thick plate. It had independent water-cooling systems, with 2 radiators. Additional components, not normally identified with the S.G.6 were an Opel 2.5 litre 4-speed gear box and Z.F. transfer box. In addition, a searchlight or spotlight appears to have been mounted on the turret.

It was 14ft 7 3/8in long, 5ft 6 7/8in wide, 5ft 1 3/8in tall, and 4ft 4in tall less the search light. The wheel centres were 8ft 2 3/8in. Track front and rear 4ft 9 3/4in. It weighed 1-ton 10-cwt without crew or a full petrol tank. The steering system was a Trippel worm and sector type.

An unsuccessful demonstration of the Schildkröte I took place before Army authorities in Berlin in the latter half of March 1942, upon which this design was cancelled and development of a second version begun. An order for two further vehicles was placed with Trippel-Werke GmbH on 14 April 1942. The characteristics of the Schildkröte II are as follows:

Mechanical layout and components as for Schildkröte I, except for steering gear. Special steering gear developed. Armour basis increased to 10mm. An unsuccessful experiment was made with paddle (bladed wheel) water propulsion and was abandoned in favour of propellor drive. It was powered by either an air-cooled Phanonomen 27 engine or an air-cooled Tatra 8-cylinder engine mounted in the rear of the vehicle. It had a 4 or 5 speed standard Opel gearbox with an additional high/low 2-speed box. The differentials were self-locking. It had all four-wheel drive, and it used a propellor in the water which was engaged by a control in the driver's compartment. A rudder was used to steer the vehicle in the water. The tyres were bullet proof, and the suspension used Helical springs and hydraulic shock absorbers. It had a maximum road speed of 50mph and in water 5.5mph. It was 4ft 7 1/8in tall excluding the turret, 6ft 2 3/4in wide, 15ft 11in long and it had a ground clearance of 10in.

The Schildkröte II was successfully demonstrated before Field Marshal Milch and his staff on 1 and 2 May 1942 and in addition in Berlin in June 1942. The vehicle demonstrated was without a turret, although it was intended to mount an open-topped octagonal turret with a 20mm M.G.151/20 gun or a 7.98mm M.G.81 gun with a coaxial 7.98mm M.G.34 machine gun. It weighed 2-tons 11-cwt (2,658kg). It had a crew of two; the driver sat in the hull and the gunner in the turret.

Work on a further development, the Schildkröte III, was started in June 1942 and was demonstrated in Berlin on 1 October 1942. The hull of this vehicle is believed to have been the subject of the drawing according to Appendix G. A brief specification listing its chief distinguishing characteristics were as follows:

10mm armour basis, Turret mounted M.G.151. S.G.6 chassis and suspension. Steering as for Schildkröte II. Opel 3.6 litre gearbox and type A auxiliary box. Tatra 75hp engine.

EDITOR'S NOTE

In 1942, the Schildkröte I, II and III prototype designs were unable to compete with the current German armoured cars of the time. Various problems plagued the project; the light construction of the chassis made the Schildkröte unable to carry heavier armour and armament. Trippel reported that it had problems in acquiring sufficient materials to start full-scale production. The Schildkröte project was cancelled by late 1942.

Front view of a Schildkröte III prototype vehicle undergoing water trials. It was armed with a 20mm M.G.151/20 gun.

Side view of a Schildkröte III prototype vehicle undergoing water trials. It was armed with a 20mm M.G.151/20 gun.

In 1943-44, Trippel once again attempted to design an armoured car suitable for military use, resulting in the Einheit Panzerspähwagen Trippel E3 (Trippel's E3 Army Unit Armoured Reconnaissance Vehicle), based on the Schildkröte III. The E3 featured a more powerful, air-cooled Tatra V-8 125hp petrol engine and armour protection ranging from 5.5mm to 14.5mm. This too failed, and in October of 1944, Waffenamt concluded that such a vehicle was unnecessary.

Bibliography

BOOKS

Jentz, Thomas L. and Louis Doyle, Hilary Louis, *Panzer Tracts No. 2-1* (Panzer Tracts, 2002)
—*Panzer Tracts No. 3-1* (Panzer Tracts, 2003)
—*Panzer Tracts No. 4* (Panzer Tracts, 2023)
—*Panzer Tracts No. 18* (Panzer Tracts, 2007)
—*Panzer Tracts No. 6-3* (Panzer Tracts, 2013)

OFFICIAL REPORTS, PAMPHLETS AND FIELD MANUALS

Experimental report on 8-ton tank Praga TNH-P, MEE Report No. A.99, 13 June 1939, W.O. File 57/tanks/2600. A=M.B. Test sheet No. A.357

Fighting Vehicle Proving Establishment Field Trials Report on German Light Tank Pz.Kw.II (ex-Libya) F.V.P.E. Field Trial Report No. F.T.679

Panzer III Poison Gas bullet attack. National Archives WO 189/2365 and WO 189/2405

Panzer IV Ausf D tank, Department of Tank Design, D.T.D Experimental F.T. 214, National Archives WO 194/181

ETO Ordnance Technical Intelligence Report No. 131, 27 January 1945. Subject: Wire-Mesh skirting on German tanks

E.100 – German Tanks, Fighting Vehicles Divisional Letter No.15 National Archives WO 194/158

Fighting Vehicle Proving Establishment Field Trial Report on Armoured Car, Autoblinda 40 (41) Italina 4x4 Performance trials. Report No. F.T.1595, D.T.D. Project No. D.E.10074, D.T.D. File No. 10/40/41. National Archives WO 194/214

Schildkröte Panzerspähwagen Light Amphibious Turtle Armoured Car, Appendix 'B' to War Office Technical Intelligence Summary No. 184 dated 23 August 1945